Disaster Risk Reduction

Methods, Approaches and Practices

Series Editor

Rajib Shaw, Keio University, Shonan Fujisawa Campus, Fujisawa, Japan

Disaster risk reduction is a process that leads to the safety of communities and nations. After the 2005 World Conference on Disaster Reduction, held in Kobe, Japan, the Hyogo Framework for Action (HFA) was adopted as a framework for risk reduction. The academic research and higher education in disaster risk reduction has made, and continues to make, a gradual shift from pure basic research to applied, implementation-oriented research. More emphasis is being given to multi-stakeholder collaboration and multi-disciplinary research. Emerging university networks in Asia, Europe, Africa, and the Americas have urged process-oriented research in the disaster risk reduction field. With this in mind, this new series will promote the output of action research on disaster risk reduction, which will be useful for a wide range of stakeholders including academicians, professionals, practitioners, and students and researchers in related fields. The series will focus on emerging needs in the risk reduction field, starting from climate change adaptation, urban ecosystem, coastal risk reduction, education for sustainable development, community-based practices, risk communication, and human security, among other areas. Through academic review, this series will encourage young researchers and practitioners to analyze field practices and link them to theory and policies with logic, data, and evidence. In this way, the series will emphasize evidence-based risk reduction methods, approaches, and practices.

Mikio Ishiwatari · Daisuke Sasaki
Editors

Financing Investment in Disaster Risk Reduction and Climate Change Adaptation

Asian Perspectives

Springer

Editors
Mikio Ishiwatari
Graduate School of Frontier Sciences
The University of Tokyo
Chiba, Japan

Daisuke Sasaki
International Research Institute of Disaster Science (IRIDeS)
Tohoku University
Sendai, Japan

ISSN 2196-4106 ISSN 2196-4114 (electronic)
Disaster Risk Reduction
ISBN 978-981-19-2923-6 ISBN 978-981-19-2924-3 (eBook)
https://doi.org/10.1007/978-981-19-2924-3

This Springer imprint is published by the registered company Springer Nature Singapore Pte Ltd.
The registered company address is: 152 Beach Road, #21-01/04 Gateway East, Singapore 189721, Singapore

Preface

It is widely recognized that investment is crucial to reducing disaster risks and adapting to the effects of climate change. However, most countries are unable to proactively invest enough in disaster risk reduction (DRR) and climate change adaptation (CCA). To promote investment in DRR and CCA, there are various issues to be overcome. For example, there is no clear estimation of finance for DRR and CCA activities at the global scale. Governments and international organizations should ensure harmony among response activities, recovery efforts, and development. Research gaps cover developing methods to weigh the costs and benefits of DRR and CCA to justify the investment.

This book has 12 chapters, which deal with various aspects of investment in DRR and CCA. The first section identifies the key issues in the Asia-Pacific region and steps to promote investment by international organizations. The second section focuses on case studies of analyzing the scale of investment in DRR and CCA, the practices of financial arrangements for DRR, and opportunities and challenges. The third section proposes innovative approaches and policies for filling the gaps in the efforts and means of securing finance for humanitarian response and recovery, and green infrastructure.

This book is designed to provide students, researchers, and practitioners with significant aspects of investment in DRR and CCA especially in the Asia-Pacific region. It will help to better understand the issues and opportunities of investments. We hope that the readers will find this book useful and can use it in their own work.

Chiba, Japan — Mikio Ishiwatari
Sendai, Japan — Daisuke Sasaki

About This Book

As the population is growing and urbanization is progressing, higher numbers of people are exposed to disaster risks, especially in the developing countries. Climate change is further worsening the impacts of existing risks and introducing new ones. It is imperative that countries need to invest more in disaster risk reduction (DRR) as well as climate change adaptation (CCA) not only to minimize their impacts but also to build resilience.

To date, various international arenas have recommended increasing investment in DRR. The Sendai Framework for DRR, adopted by the UN member states in 2015 during the Third UN World Conference on DRR, emphasizes investment as a priority for decreasing disaster risks and losses. The Yangon and Kumamoto declarations at Asia-Pacific water summits set the goal of doubling investment to address water-related disasters. However, most countries are not able to proactively invest enough in DRR. Understanding the current scale and estimate of investment and its effects is crucial for promoting DRR investment, but such information and estimates are rarely available.

This book examines the current investment trends and issues in DRR and CCA. Based on specific case studies, field data and evidence, the book identifies challenges in increasing investment and recommends various investment policies and innovative approaches to sustainable progress towards a resilient future.

Contents

About the Editors

Dr. Mikio Ishiwatari is a senior advisor on disaster management and water resources management at the Japan International Cooperation Agency, and a visiting professor at the Graduate School of Frontier Sciences, The University of Tokyo. He has been engaged in projects and research work on disaster risk reduction, climate change adaptation, and water issues. He led the formulation of the Japanese assistance policies of climate change adaptation and community-based disaster management.

At the World Bank he worked as a senior disaster risk management specialist and published Learning from Megadisaster: Lessons from the Great East Japan Earthquake. He worked in various positions at the Ministry of Land, Infrastructure, and Transport, Japan, for 17 years, and was responsible for research and technology development as a senior deputy director for river technology and information. He also worked as an urban development specialist at the Asian Development Bank. He holds a Ph.D. in international studies and an M.Sc. in urban engineering from The University of Tokyo.

Dr. Daisuke Sasaki is an associate professor of the International Research Institute of Disaster Science (IRIDeS) at Tohoku University in Sendai, Japan. He holds a Ph.D. (2015) in international studies and also has expertise in disaster science. He has been continuously conducting research focusing on the factors that might hinder the resolution of regional issues from the perspective of international cooperation both qualitatively and quantitatively, and also engaged in plenty of international collaborative research with several universities in the Asia-Pacific region.

Dr. Sasaki has published a number of articles in refereed journals including the International Journal of Disaster Risk Reduction (IJDRR), and he has served as a guest editor for several special issues of the Journal of Disaster Research (JDR).

Contributors

Fujikura Ryo Faculty of Sustainability Studies, Hosei University, Chiyodaku, Tokyo, Japan

Ishiwatari Mikio Graduate School of Frontier Sciences, The University of Tokyo, Chiba, Japan

Iuchi Kanako International Research Institute of Disaster Science (IRIDeS), Tohoku University, Aramaki, Aoba-ku, Sendai, Miyagi, Japan

Jibiki Yasuhito International Research Institute of Disaster Science (IRIDeS), Tohoku University, Aramaki, Aoba-ku, Sendai, Miyagi, Japan;
Center for Integrated Disaster Information Research (CIDIR), The University of Tokyo, Bunkyo-ku, Tokyo, Japan

Katsube Tsukasa Japan International Cooperation Agency, Tokyo, Japan

Mizutani Daijiro Department of Civil and Environmental Engineering, Tohoku University, Miyagi, Japan

Nakamura Keigo Japan Riverfront Research Center, Tokyo, Japan

Pelupessy Dicky C. Faculty of Psychology, Universitas Indonesia, Depok, West Java, Indonesia

Sakamoto Masashi International and Regional Cooperation Office, International Research Institute of Disaster Science (IRIDeS), Tohoku University, Sendai, Japan;
Disaster Prevention Planning Dept, Land Infrastructure Div, Pacific Consultants Co., Ltd, Tokyo, Japan

Sasaki Daisuke International Research Institute of Disaster Science (IRIDeS), Tohoku University, Sendai, Miyagi, Japan

Shiiba Nagisa Institute for Global Environmental Studies, Kanagawa, Japan

Shiohara Kaori The University of Tokyo, Tokyo, Japan

Part I
Understanding the Context: Investing in Disaster Risk Reduction and Climate Change Adaptation in Asia

Chapter 1
Financing Investment in Disaster Risk Reduction and Climate Change Adaptation: Opportunities and Challenges

Mikio Ishiwatari and Daisuke Sasaki

Abstract Asia, a rapidly emerging global economic power, is vulnerable to natural disasters. Its annualized flood-related losses, amounting to 53 billion USD and approximately 16,000 human lives, are expected to increase with further climate change. Several international organizations have recommended increasing investment in disaster risk reduction (DRR) and climate change adaptation (CCA). Developing Asian economies' investment of 33.6 billion USD (0.21% of their GDP) in flood protection in 2015 is projected to reach 100 billion USD annually by 2030. Official development assistance (ODA) in DRR—5 billion USD/year from 1991 to 2010—increased by 2.6 times to 13.3 billion USD/year from 2011 to 2020. Simultaneously, investment in risk prevention or reduction decreased from 680 to 550 million USD/year (a decrease from 12.7 to 4.1% as the share in DRR assistance). This book collates the existing data, research findings, conceptual works of the scholars and senior officials of disaster management organizations who directly manage or investigate natural disasters in the Asia-Pacific, and practical cases of investment in DRR and CCA. This chapter examines the international initiatives for promoting investment in DRR and CCA and the current and projected funding for DRR and CCA to construct the background for this compendium and outlines its chapters.

Keywords Asia-pacific · Emergency response · Flood · Japan · Official development assistance

M. Ishiwatari (✉)
Graduate School of Frontier Sciences, The University of Tokyo, Chiba, Japan
e-mail: ishiwatari.mikio@jica.go.jp

D. Sasaki
International Research Institute of Disaster Science (IRIDeS), Tohoku University, Sendai, Miyagi, Japan
e-mail: dsasaki@irides.tohoku.ac.jp

M. Ishiwatari and D. Sasaki (eds.), *Financing Investment in Disaster Risk Reduction and Climate Change Adaptation*, Disaster Risk Reduction,
https://doi.org/10.1007/978-981-19-2924-3_1

1.1 Introduction

While Asia is rapidly emerging as an economic center of the world, it is also the region most vulnerable to natural disasters. Major cities in this region are located in low-lying areas that are vulnerable to floods and inundation by the sea. The loss of human life is exacerbated by the inflation of the urban population caused by continuing migration from rural areas. The Asian Development Bank (ADB 2016, 2020) estimated the annualized economic loss by water-related disasters including droughts at 53 billion USD and human losses at approximately 16,000 lives per year. The risk of disasters in this region is expected to increase because of climatic and socio-economic changes.

This book focuses on investments in disaster risk reduction (DRR) and climate change adaptation (CCA). Commonalities between DRR and CCA lie within the management of hydro-meteorological hazards, such as floods, landslides, and storms (Coninx et al. 2016). DRR involves managing geophysical hazards, while CCA involves non-disaster events that are affected by a changing climate (Fig. 1.1).

Investment is crucial to reducing disaster risks and adapting to the effects of climate change (Ishiwatari 2021b). The importance of investing in DRR is widely acknowledged (Ishiwatari and Surjan 2019; UNISDR 2015). However, Mizutori (2020), Special Representative of the United Nations Secretary-General for Disaster Risk Reduction, stressed that "financing such investments has so far proved to be difficult." Most countries are unable to proactively invest enough in DRR (Kawasaki and Rhyner 2018).

This chapter summarizes the layout of the book and highlights the major findings of each chapter. It presents an overview of the key issues related to investment in DRR and CCA in Asia and reviews the international initiatives undertaken for promoting

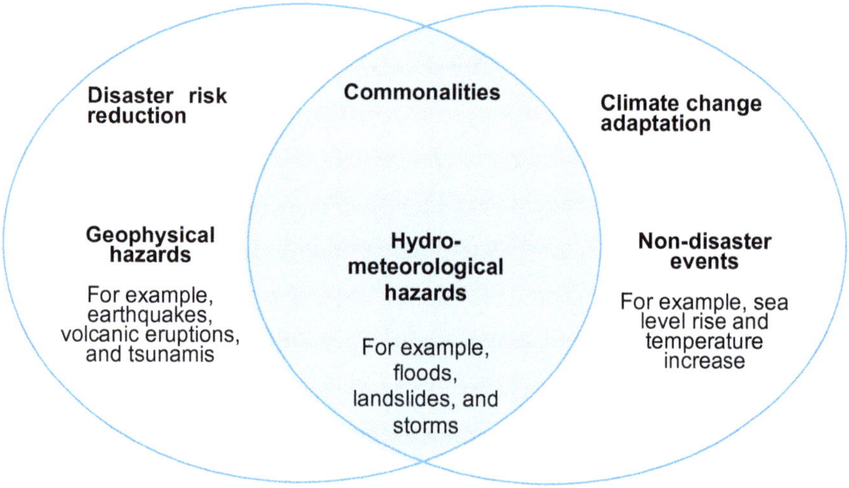

Fig. 1.1 Commonalities of DRR and CCA. Reproduced from Coninx et al. (2016)

such investment. It estimates the current and required scales of investment in flood protection and reviews the trends in ODA in DRR.

1.2 International Initiatives for Promoting DRR and CCA Investment

To date, various international organizations have recommended the enhancement of investment in DRR and CCA. In 2015, the Sendai Framework for DRR (SFDRR) highlighted the importance of investing in DRR. The framework emphasizes as its principle that disaster-risk-informed investments are more cost-effective than post-disaster response and recovery and that such investments contribute to sustainable development. Further, it emphasizes public and private sector investment in hardware and software measures as a priority action to reduce disaster risks. The UN member states adopted this framework during the Third UN World Conference on DRR in Sendai in 2015 and laid down guidelines for the member states to strengthen DRR. Investment in DRR can promote innovation, accelerate growth, create job opportunities, and strengthen the resilience of society. The framework stresses on investing in structures and protecting crucial facilities, cultural heritage, and environment, and strengthening insurance and safety net programs (UNISDR 2015). This framework was in stark contrast with the Hyogo Framework for Action adopted at the Second UN World Conference on DRR convened in 2005 at Kobe, Hyogo. Though this was the predecessor of the SFDRR, it did not make an unambiguous call for investment in DRR (UNISDR 2005).

The Paris Agreement in 2015, which is an international treaty on climate change, reaffirmed the commitment of developed countries to invest in mitigating the effects of climate change and to help developing countries offset the retarding effect of adapting to climate change on their economic development. Climate finance can help developing countries enhance their capacity to face the disasters caused by climate change (Ishiwatari 2021a; Khan et al. 2020). Through the Cancún agreement adopted at the UN Climate Change Conference (COP16) held in Cancún, Mexico in 2010, it was decided to establish the Green Climate Fund (GCF) to support developing countries. The agreement also aimed to achieve a 50:50 distribution of the fund between mitigation and adaptation measures. It was also decided that at least 50% of adaptation funds should go to particularly vulnerable countries, such as the least developed countries; small, developing island countries; and African countries (GCF 2021). At the Climate Action Summit in 2019, participating countries, international organizations, and the private sector committed themselves to enhance resilience and adaptation. The purpose of resilience initiatives includes strengthening the market for mobilizing private investment in climate-resilient infrastructure, which is mainly invested by the public sector (United Nations 2019).

Other international initiatives have supported financing of DRR and CCA following the Sendai Framework and Paris Agreement in 2015. The Yangon Declaration—The Pathway Forward—set the goal of doubling investment in projects to reduce risks of water-related disasters and increase water security in the Asia-Pacific region. The declaration was adopted at the Third Asia-Pacific Water Summit in 2017 by the heads of states and ministers, who are responsible for water issues, and other leaders (Asia Pacific Water Forum 2017).

The High-Level Panel on Water (2018) recommended doubling investment in water-related DRR within the next 5 years. This panel has been established jointly by the United Nations and the World Bank (WB) in 2016 and consists of 11 sitting heads of states and governments and one special advisor. The panel serves as the network to mobilize political will for reducing water-related disaster risks. The panel recommended changing the allocation of international financial assistance from the current 10 and 90% between disaster reduction/preparedness and emergency response/rehabilitation, respectively, to a more balanced allocation first and eventually to 90:10% (HELP 2019).

1.3 Current Scale and Estimate of Demand for DRR and CCA

To promote investment in DRR and CCA, it is important to know the current scale of investment. However, such information is rarely available because of the limited availability of budget data for DRR and CCA. This section reviews the recent studies on estimating investment in DRR.

1.3.1 Current Scale of Investment in Flood Protection

Ishiwatari (2019) collected investment data of nine flood-prone developing economies and three high-income economies in Asia. The nine developing economies, namely, Bangladesh, the People's Republic China (PRC), India, Indonesia, Malaysia, Pakistan, the Philippines, Thailand, and Vietnam may be regarded as major flood-prone Asian economies. These nine developing economies account for over 90% of the total population and the total GDP of all the developing economies in Asia. The high-income economies included in the study were Japan, the Republic of Korea, and Taiwan. The budget data of Myanmar are not available.

Nine major developing economies in Asia invested USD33.6 billion or 0.21% of their total GDP in flood protection, (Table 1.1). This amount accounted for approximately 4% of total investment in infrastructure (Ishiwatari and Sasaki 2020). Total investment in flood protection by all the countries in Asia is estimated at over USD50

Table 1.1 Estimated demand for investment in flood protection and climate change adaptation in developing Asia

Economies	Investment in flood protection in 2015	Estimated demand for investment (average from 2016 to 2030)		
	Amount USD billion per year, 2015 price (Share of GDP (%))	Amount in flood protection USD billion per year, 2015 price (Share of GDP (%))	Amount in CCA USD billion per year, 2015 price	Total Amount USD billion per year, 2015 price
PRC	29.9 (0.27)	68.7 (0.38)	2.2	70.9
India	1.5 (0.07)	13.4 (0.35)	1.1	14.5
Philippines	1.1 (0.38)	3.0 (0.62)	0.1	3.1
Indonesia	0.5 (0.06)	3.1 (0.23)	0.2	3.3
Thailand	0.5 (0,12)	1.6 (0.29)	0.1	1.7
Vietnam	0.1 (0,05)	0.6 (0.2)	0.03	0.63
Malaysia	0.03 (0.01)	1.0 (0.23)	0.02	1.02
Bangladesh	0.007 (0.004)	2.1 (0.59)	0.2	2,2
Pakistan	0.009 (0.004)	0.9 (0.24)	0.01	0.91
Total for developing economies	**33.6 (0.21)**	**94.5 (0.36)**	**3.9**	**98.4**
Japan (2014)	17.5 (0.39)	NA	NA	NA
Republic of Korea	2.0 (0.13)	NA	NA	NA
Taiwan	0.3 (0.06)	NA	NA	NA
Total for developed economies	**19.8**	**NA**	**NA**	**NA**
Total	**53.4 (0.24)**	**NA**	**NA**	**NA**

Source Modified after Ishiwatari (2020) and Ishiwatari and Sasaki (2020)

billion, or 0.24% of the total GDP. The PRC, the Philippines, Thailand, Japan, and Taiwan have been investing more than 0.1% of GDP in flood protection.

Investment in flood protection in Asia is nearly 15 times as much as in Europe and 25 times of that in the US. Investments in Europe and the US are estimated at USD3.6 billion per year and USD2.2 billion per year, respectively (Organisation for Economic Co-operation and Development (OECD) 2020; Multi-Hazard Mitigation Council 2019; USACE 2017).

Figure 1.2 shows the trends in investment. The PRC began increasing the flood protection budget in the late 1990s and increased it by over six times from 1996 to 2006 and over five times from 2006 to 2016. The Philippines increased its flood protection budget by over eight times from 2008 to 2016. India increased the budget by four times during the last decade. These economies have increased their budgets

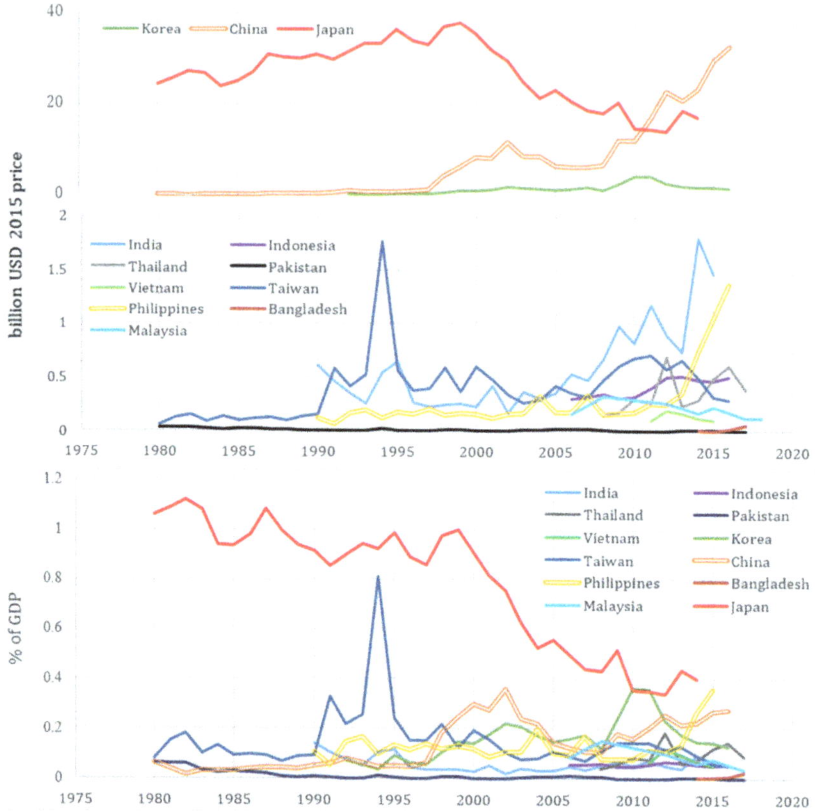

Fig. 1.2 Trends in Flood Protection Investment in 12 Major Asian Economies. *Source* Ishiwatari (2019)

to prevent the increasing economic damage caused by floods. Other economies including the high-income economies of Japan, Taiwan, and the Republic of Korea have not consistently increased their budgets.

It may be observed that the scale and effect of the investments in flood protection have kept pace with the economic development (Table 1.2). Low-income economies

Table 1.2 Budgeting for flood protection and flood damage according to development stages

Development stage (GDP per capita USD)	Budget	Damage
Low-income (less than 1,000)	Not increased	Increased
Lower-middle-income (1,000–4,000)	Increased	Not increased
Upper-middle-income (4,000–12,000)	Increased	Not increased

Source Data fromIshiwatari (2019)

can invest limited amounts in flood protection, and economic damage increases as the economy develops and the additional investment that brought economic development suffers damage. Lower-middle-income economies start investing more in flood protection and the damage does not increase. Upper-middle-income economies continue to increase investment till the damage level stabilizes. However, if flooding exceeds the designed safety level of the protection control structure, it can cause damage on an unprecedented scale, as was caused by the floods in Thailand during 2011 (Ishiwatari 2019).

Ishiwatari and Sasaki (2021a) applied multiple regression models to analyze the relationships among investment in flood protection, flood damage, and socio-economic factors in developing economies of Asia. It was found that a few years after flood damage increased, governments increased their budgets for flood protection. These economies increased their budgets further when per capita GDP and population density increased. There is no clear relationship between the sectoral composition of the economy, urbanization, and investment in flood protection. From these findings, it can be interpreted that governments tend to allocate less than the necessary budgets for flood protection during normal times and are compelled to incur the expenditure on post-event recovery and prevention for future floods. This is mainly because it is not easy for policymakers and the general public to understand flood risks and prepare for them (World Meteorological Organization 2012; OECD 2016).

Investment in CCA is increasing and has reached USD 30 billion in 2017–2018. It accounted for 5% of climate finance (Buchner et al. 2019). Multilateral development banks allocated around 20 million USD, which was less than 20% of climate finance committed for CCA activities in 2018 (MDB 2019).

1.3.2 Estimation of Investment in Flood Protection and CCA

Since the current scales of investment in CCA and DRR are unclear, estimating the demand for finance is a challenge. This estimation has been attempted in very few studies.

The Global Commission on Adaptation (2019) estimated adaptation costs in early warning systems, climate-resilient infrastructure, improved dryland agriculture. The Commission put the cost at 1.8 trillion USD globally between 2020 and 2030, or 180 billion USD per year. This investment could generate total benefits of 7.1 trillion USD during the decade. The United Nations Environment Programme (2021) estimated the current annual adaptation costs in developing countries alone at 70 billion USD, 140–300 billion USD in 2030, and 280–500 billion USD in 2050.

Ishiwatari and Sasaki (2020) developed the multi-regression model of investment in flood protection and estimated that, of the total amount of 98.4 billion USD invested annually by the developing economies of Asia in flood-related works during the period of 2016–2030, 94.5 billion USD were invested in flood protection infrastructure, and 3.9 billion USD were spent on dealing with climate change effects

(Table 1.1). The financing gap, which is future needs minus current investment, is around 61 billion USD annually in flood protection infrastructure. The gap widens to 65 billion USD with the additional finance needed to deal with climate change effects. This amounts to about 0.24% of GDP. The cumulative net benefit from this investment, which is benefit minus cost, would become positive in 2028 and reach 263 billion USD in 2030 (Ishiwatari and Sasaki 2021b).

Rozenberg and Fay (2019) estimated that low- and middle-income economies would invest between 0.46 and 0.52% of their GDP in flood and coastal protection annually by 2030 in the East and South Asia and Pacific regions. They conceptualized and studied various scenarios and estimated future flood protection investments based on (i) various levels of risk tolerance, (ii) various socioeconomic scenarios, (iii) various per unit investment costs, and (iv) various climate change scenarios. The investment, which includes new capital and maintenance, aims at minimizing overall costs of protection and compensating for residual flood damages. They proposed a preferred scenario with ambitious goals and high-efficiency measures that would require an investment of USD103 billion, that is, 0.32% of GDP.

1.3.3 Trends in Official Development Assistance for DRR

Official development assistance for DRR increased 2.6 times from 5 billion USD/year during the period of 1991–2010 to 13.3 billion USD/year in 2011–2020 (Fig. 1.3). Over this period, the share of DRR assistance in total development assistance increased from 3.5 to 11%. However, during this period, financing for ex-ante measures to reduce and prevent risks before disaster occurrence declined from $680 million/year to $550 million/year, and the share of DRR assistance in it declined from 12.7 to 4.1% (Kellett and Caravani 2013; UNDRR 2021).

The three major donors for DRR, WB, ADB, and Japan provided financial assistance for DRR amounting to 6 billion USD/year. The three donors provided over

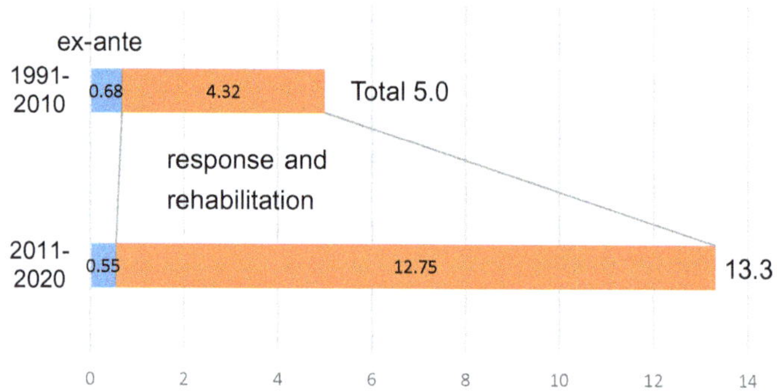

Fig. 1.3 Official development assistance for DRR. Numbers are billions of USD. *Source* Data collected from Kellett and Caravani (2013) and UNDRR (2021)

70% of assistance for DRR between 1991 and 2010 (Kellett and Caravani 2013). WB provided 4.5 billion USD for DRR in 2020, and Japan provided 850 million USD in 2019 (Ministry of Foreign Affairs 2021; World Bank 2021). ADB provided 170 million USD annually on average for non-emergency, ordinarily projects between 2014 and 2018, and 480 million USD annually on average as emergency assistance to cope with post-disaster situations between 2004 and 2018 (ADB 2019).

The amount of international humanitarian assistance worldwide reached nearly 31 billion USD in 2018 and 2019, and 30.9 billion USD in 2020. This included 6.7 billion USD from the private sector (Development Initiatives 2021). This amount was for responding to the effects of both man-made crises and natural disasters and, therefore, cannot be directly compared with the ODA in DRR of 13.3 billion USD. Distinguishing assistance given to deal with natural disasters from other humanitarian assistance is difficult because these crises quite often come together to form complex disasters. Among 40 countries that needed aid to meet the humanitarian needs of over 1 million people, 33 countries experienced natural disasters, 27 experienced conflicts, and 27 experienced displacements of population.

1.4 Issues Affecting the Promotion of Investment in DRR and CCA

This section examines the key issues dealt with in the chapters of this book. These issues span a wide range of areas of data, policies, and practical approaches.

1.4.1 Definition and Data Collection

The database of investments in DRR and CCA needs to be further strengthened. The data on the investment by national governments and ODA are available, though incomplete, as discussed in the preceding sections. However, collecting data on investments made by local governments, local communities, and the private sector is challenging (Shiohara 2022). While the amounts spent on mitigation of the effects of climate change by the private sector are reported, CCA by the private sector is not fully recorded (Fujikura 2022). The private flows of climate change-related finance are placed tentatively at 0.1% of the total financing of CCA because of the limited availability of data (CPI 2019). Further, the data of investment in specific areas, such as coastal zones, are not clearly categorized (Shiiba 2022).

1.4.2 Harmonization of Response Activities, Recovery Efforts, and Development

When a disaster strikes, funds flow into disaster-affected areas to aid response and recovery efforts. These funds are precious resources and help to quicken the recovery and create a resilient and sustainable society in developing countries, where resources for development are limited. Given these positive effects of aid, relief and recovery activities need to be linked to long-term development (Gómez and Kawaguchi 2018; Ishiwatari 2018; JIbiki et al. 2022; Katsube 2022).

1.4.3 Justification of Investment in DRR and CCA

Policy- and decision-makers need hard facts on which to base investment policies and promote projects for DRR and CCA. Research can provide evidence for decision- and policy-making. However, there are various gaps in research (Sasaki and Ishiwatari 2022). Therefore, every aspect of disaster has not yet been discovered. It is not clear which factors influence people's perception of investments (Pelupessy et al. 2022). Cost-benefit analysis is widely used to examine the efficacy of DRR projects and can provide necessary information for policy and project considerations. However, estimating all the benefits is challenging. For example, putting a price on human life that is lost or saved is controversial (Sakamoto et al. 2022). There is a need to develop methods for evaluating indirect benefits, such as mitigating interruptions in supply chains and increasing land prices. Furthermore, lack of regular maintenance can degrade infrastructures and make them vulnerable to disasters (Chang 2003). Establishing a system for the maintenance of infrastructure is a challenge in developing countries and industrialized countries (Durango and Madanat 2002; Hemme 2015). In this context, optimizing the maintenance of infrastructure from the DRR perspective will be of prime importance in the coming decades (Muzutani 2022).

1.4.4 Investment in Green Infrastructure as Innovative Measures

Green infrastructure uses natural functions that serve the purpose of reducing the risks of disaster and providing numerous benefits, such as climate change mitigation, greater biodiversity, better water quality, and means of recreation (Nakamura 2022). In the US, wetlands reduced the extent of the damage that Hurricane Sandy could have caused by about 1% in 2012 (Narayan et al. 2017). Coral reefs reduce by one-tenth to one-fifth the inundation-caused damage to high-value real estate on wealthy coastal areas of Hawaii, Florida, and Puerto Rico (Reguero et al. 2018). Green infrastructure is more cost-effective than the gray infrastructure made from concrete and steel.

1.5 About the Book

The book is in three parts. Part I gives an overview of the key issues to be considered while investing in DRR and CCA, and steps to promote investment by international organizations. Part II focuses on case studies of Japan, Indonesia, and local governments in the Philippines, and infrastructure and coastal resilience. Part III proposes innovative approaches and policies for filling the gaps in the efforts and means of securing finance for humanitarian response and recovery, and green infrastructure. Further, it synthesizes the major implications based on the key findings of the previous chapters and offers a pathway to attract investment in the Asian Region.

1.5.1 Part I Understanding the Context: Investing in Disaster Risk Reduction and Climate Change Adaptation in Asia

Chapter 2 describes CCA financing and argues the continuing need for widely accepted standards to measure the required amount of adaptation finance. The amount of CCA financing from developed countries to developing countries is estimated to have been USD30 billion in 2017–2018. However, this may be an overestimate because the CCA is not clearly defined. To fill the financial gap in CCA, it is expected that innovative modalities would be expanded. New financing mechanisms, such as Climate Resilience Bonds and catastrophe bonds are engaging the attention of developing countries as the vehicles to introduce private capital into the development of resilient infrastructure.

Chapter 3 reviews the literature on DRR investment following the introduction of SFDRR in 2015. The review revealed that the nomenclature used in the literature produced by multilateral development banks could indicate the direction for further research. WB and ADB provide the new concepts of the triple dividend of resilience and women-focused investment. The triple dividend of resilience covers (i) avoiding disaster losses, (ii) stimulating economic activities, and (iii) providing socio-economic co-benefits and environment. Some academic papers provide evidence for these new concepts discussed by WB and ADB.

1.5.2 Part II Case Studies

Chapter 4 explores people's perception of DRR investment in flood management in Jakarta. "The sense of resilience" significantly affects people's investment needs. Among factors of "the sense of resilience," "place attachment," and "collective efficacy" can affect investment needs. There is no clear relationship between "the sense of community" and "the sense of resilience."

Chapter 5 describes factors influencing investment in CCA at the local level in the Philippines. For the local governments, the share of CCA finance in the investment does not increase as the scale of local governments' budgets increase. The amount invested in CCA does not relate to flood risks. Local governments that receive funding from public works departments of the national government spend more of the CCA budgets. This is because constructing the resilient infrastructure of highways, flood control structures, and agriculture development are categorized as CCA investments. It was also found that the political connections of the local government with the national government could increase CCA investment.

Chapter 6 focuses on optimizing investments in resilient infrastructure taking into account estimating maintenance costs warranted by the level of disaster risks. The chapter draws upon a case study of the maintenance cost of a bridge damaged in an earthquake in Japan. As the probability of disasters increases, routine maintenance needs to be strengthened to prevent the deterioration of structures and reduce disaster damage. Thus, assessing the state of structural deterioration and risks posed by disasters is crucial for managing infrastructure. Disaster management and maintenance should be integrated to reduce costs from the perspective of managing life cycle costs.

Chapter 7 presents multilateral aid for CCA in the regions on ocean and sea coasts in the Asia Pacific. Climate change will damage the coastal environment and the resources in it, and, consequently, economic activities including tourism will suffer losses. CCA in ocean and coastal areas can be regarded as a crucial component for blue finance, which supports the sustainable use and conservation of the oceans. The flow of finance into CCA increased from USD 850 million in 2010 to more than USD 4.3 billion in 2020. Of this amount, USD 2.5 billion was used for developing infrastructure. Out of the total number of projects, ecosystem conservation accounts for 32%, followed by urban infrastructure, and fisheries and aquaculture at 21% each.

1.5.3 Part III Policy for Bridging Gaps

Chapter 8 analyzes funding for humanitarian response by focusing on early recovery from disaster. The international community regarded activity undertaken to ensure early recovery as non-life-saving, but the recent trend has been to emphasize the connection of response activities with recovery and development efforts. To describe the assessment processes and financing mechanisms of humanitarian response, this chapter uses four disasters as examples, namely, the disaster caused by Typhoon Haiyan 2013 in the Philippines, Cyclone Pam 2015 in Vanuatu, Gorkha Earthquake 2015 in Nepal, and Sulawesi Earthquake 2018 in Indonesia. It was found that the need for early recovery accounts for the limited amount of 0.04–0.06% of the total need for humanitarian assistance. The means of securing funds vary according to the nature of the disaster.

Chapter 9 describes the weak linkages between post-disaster needs assessment (PDNA) and investment in DRR and CCA. Government organizations, in collaboration with foreign and international organizations, conduct PDNA to assess financing needs for disaster recovery activities covering numerous sectors, such as infrastructure, shelter, and livelihoods. PDNA should include investment to prepare for future disasters in addition to rehabilitating the damaged infrastructure. However, case studies in the Philippines and Indonesia show limited coordination among the offices of disaster management and climate change. Laws, ordinances, and guidelines related to PDNA do not adequately state the investment in DRR and CCA.

Chapter 10 describes nature-based solutions, focusing on the case of Japan. In Japan, nature-based solutions are often referred to as green infrastructure. Efforts for building green infrastructure have been expanding in Japan since the Great East Japan Earthquake and Tsunami in 2011. A combination of severe water disasters caused by climate change and people's desire for well-being has drawn attention to green infrastructure. Japan's Ministry of Land, Infrastructure, Transport and Tourism (MLIT) established the Green Infrastructure Public-Private Partnership Platform in 2020 to promote green infrastructure in collaboration with the private sector.

Chapter 11 reviews the evolution process of the methods for cost-effectiveness analysis of the flood protection projects in Japan and cost-benefit analysis of overseas flood protection projects supported by the Japanese official development assistance. Cost-effectiveness analysis is crucial in prioritizing and justifying DRR projects. Japan has improved the analysis methods based on the datasets of flood damage. However, it has not been able to develop methods that quantify some benefits, such as mitigating psychological damage or saving human lives. The case studies of assistance projects show difficulties in estimating project benefits in developing countries because of the limited availability of data on geography and the disaster-caused damage.

Chapter 12 draws together the major findings and synthesizes the lessons drawn from the preceding chapters. It proposes policies and innovative approaches of investment to achieve sustainable progress toward a resilient future in the Asian region. It concludes by pointing out the directions that future research should take to fill the observed research gaps.

Acknowledgements This study was supported by the collaborative research project of the International Research Institute of Disaster Science, Tohoku University. This study was also partially supported by the Japan Society for the Promotion of Science KAKENHI, Grant Number 21H03680.

References

Asian Development Bank (ADB) (2016) Asian water development outlook 2016: strengthening water security in Asia and the Pacific: advancing water security across Asia and the Pacific. ADB, Mandaluyong

ADB (2019) Review of the 2004 disaster and emergency assistance policy. ADB, Mandaluyong

ADB (2020) Asian water development outlook 2020: advancing water security across Asia and the Pacific. ADB, Mandaluyong
Asia Pacific Water Forum (2017) Yangon declaration: the pathway forward. https://apwf.org/apwf_wp/wp-content/uploads/2017/12/Yangon-Declaration.pdf
Buchner B et al (2019) Global landscape of climate finance 2019. Climate Policy Initiative, London. https://climatepolicyinitiative.org/publication/global-climate-finance-2019/
Chang SE (2003) Evaluating disaster mitigations: methodology for urban infrastructure systems. Nat Hazards Rev 4(4):186–196
Coninx I et al (2016) Evolving issues brief 2016. https://www.placard-network.eu/wp-content/PDFs/Evolving-issues-brief-2016.pdf
CPI (2019) Global landscape of climate finance 2019. Climate Policy Initiative, London
Development Initiatives (2021) Global Humanitarian Assistance Report 2021. https://devinit.org/resources/global-humanitarian-assistance-report-2021/
Durango PL, Madanat SM (2002) Optimal maintenance and repair policies in infrastructure management under uncertain facility deterioration rates: an adaptive control approach. Transp Res Part A Policy Pract 36(9):763–778
Fujikura R (2022) Financing in climate change adaptation. In: Ishiwatari M, Sasaki D (eds) Financing investment in disaster risk reduction and climate change adaptation: Asian Perspectives. Springer, Tokyo
Global Commission on Adaptation (2019) Adapt now: a global call for leadership on climate resilience. https://gca.org/wp-content/uploads/2019/09/GlobalCommission_Report_FINAL.pdf
Gómez OA, Kawaguchi C (2018) A theory for the continuum: Multiple approaches to humanitarian crises management. In: Hanatani A, Gómez OA, Kawaguchi C (eds) Crisis management beyond the humanitarian-development nexus. Routledge, pp 15–25
Green Climate Fund (2021) GCF in brief: adaptation planning. https://www.greenclimate.fund/document/gcf-brief-adaptation-planning
Hemme K (2015) Critical infrastructure protection: maintenance is national security. J Strateg Secur 8(3):25–39
High-level Experts and Leaders Panel on Water and Disasters (HELP) (2018) Making every drop count, high-level panel on water. Outcome document. https://sustainabledevelopment.un.org/content/documents/17825HLPW_Outcome.pdf
HELP (2019) Principles on investment and financing for water-related disaster risk reduction. http://www.wateranddisaster.org/cms310261/wp-content/uploads/2019/07/HELP-Principles-Full-Final-Printing.pdf
Ishiwatari M (2018) How can recovery be linked with long-term development?: the case of Indonesia. In: Hanatani A, Gómez OA, Kawaguchi C (eds) Crisis management beyond the humanitarian-development nexus. Routledge, pp 162–184
Ishiwatari M (2019) Investing in disaster risk reduction: scale and effect of investment in flood protection in Asia. UNDRR, Geneva
Ishiwatari M (2021a) Disaster risk reduction. In: Lackner M, Sajjadi B, Chen WY (eds) handbook of climate change mitigation and adaptation. Springer, New York. https://doi.org/10.1007/978-1-4614-6431-0_147-1
Ishiwatari M (2021b) Effectiveness of investing in flood protection in metropolitan areas: lessons from 2019 Typhoon Hagibis in Japan. Int J Disaster Resil Built Environ 13(1):89–98
Ishiwatari M, Sasaki D (2020) Bridging the gaps in infrastructure investment for flood protection in Asia. JICA Research Institute, Tokyo
Ishiwatari M, Sasaki D (2021a) Investing in flood protection in Asia: an empirical study focusing on the relationship between investment and damage. Progr Disaster Sci 100197. https://doi.org/10.1016/j.pdisas.2021a.100197
Ishiwatari M, Sasaki D (2021b) Investments in flood protection: trends in flood damage and protection in growing Asian economies. JICA Research Institute, Tokyo

Ishiwatari M, Surjan A (2019) Good enough today is not enough tomorrow: challenges of increasing investments in disaster risk reduction and climate change adaptation. Progr Disaster Sci 1. https://doi.org/10.1016/j.pdisas.2019.100007
Jibiki Y et al (2022) Is the post-disaster needs assessment important to investment in disaster risk reduction and climate change adaptation? lessons from Indonesia and the Philippines based on literature review. In: Ishiwatari M, Sasaki D (eds) Financing investment in disaster risk reduction and climate change adaptation: Asian perspectives. Springer, Tokyo
Katsube T (2022) Early Recovery in Humanitarian Response-in cases of Four Disasters in Asia Pacific Region. In Ishiwatari M, Sasaki D (eds) Financing investment in disaster risk reduction and climate change adaptation: Asian perspectives. Springer, Tokyo
Kawasaki A, Rhyner J (2018) Investing in disaster risk reduction for resilience: roles of science, technology, and education. J Disaster Emerg Res 13(7):1181–1186
Kellett J, Caravani A (2013) Financing disaster risk reduction: a 20 year story of international aid. ODI, London
Khan M et al (2020) Twenty-five years of adaptation finance through a climate justice lens. Clim Change 161(2):251–269
Ministry of Foreign Affairs (2021) White paper on development cooperation 2020. Ministry of Foreign Affairs, Tokyo
Mizutani D (2022) Financing Infrastructure of DRR: Possibility of integrating investments in maintenance and DRR. In: Ishiwatari M, Sasaki D (eds) Financing investment in disaster risk reduction and climate change adaptation: Asian perspectives. Springer, Tokyo
Mizutori M (2020) Reflections on the Sendai framework for disaster risk reduction: five years since its adoption. Int J Disaster Risk Sci 11:147–151. https://doi.org/10.1007/s13753-020-00261-2
Multi-Hazard Mitigation Council (2019) Natural hazard mitigation saves 2019 report. National Institute of Building Sciences, Washington DC
Multilateral Development Banks (MDB) (2019) Joint report on Multilateral Development Banks' climate finance 2018. http://www.ebrd.com/2018-joint-report-on-mdbs-climate-finance
Nakamura K (2022) Nature-based solution in Japan: practice of river restoration. In: Ishiwatari M, Sasaki D (eds) Financing investment in disaster risk reduction and climate change adaptation: Asian perspectives. Springer, Tokyo
Narayan S et al (2017) The value of coastal wetlands for flood damage reduction in the Northeastern USA. Sci Rep 7:9463. https://doi.org/10.1038/s41598-017-09269-z
Organisation for Economic Co-operation and Development (OECD) (2016) Financial management of flood risk. OECD Publishing, Paris. https://doi.org/10.1787/9789264257689-en
OECD (2020) Financing water supply, sanitation and flood protection: challenges in EU member states and policy options. OECD Studies on Water, OECD Publishing, Paris
Pelupessy D et al (2022) Exploring people's perception of disaster risk reduction investment for flood management: a case from Jakarta floods in Indonesia. In: Ishiwatari M, Sasaki D (eds) Financing investment in disaster risk reduction and climate change adaptation: Asian perspectives. Springer, Tokyo
Reguero BG et al (2018) Comparing the cost-effectiveness of nature-based and coastal adaptation: a case study from the gulf coast of the United States. PLoS ONE 13(4):e0192132. https://doi.org/10.1371/journal.pone.0192132
Rozenberg J, Fay M (eds) (2019) Beyond the gap: how countries can afford the infrastructure they need while protecting the planet. Sustain Infrastr Ser (World Bank, Washington, DC)
Sakamoto M et al (2022) Estimating the benefits of flood protection: reviewing the transition of the cost-benefit analysis manual in Japan and the implementation status of the international flood protection project. In: Ishiwatari M, Sasaki D (eds) Financing investment in disaster risk reduction and climate change adaptation: Asian perspectives. Springer, Tokyo
Sasaki D, Ishiwatari M (2022) Recent trends in disaster risk reduction investments: a literature review. In: Ishiwatari M, Sasaki D (eds) Financing investment in disaster risk reduction and climate change adaptation: Asian perspectives. Springer, Tokyo

Shiiba N (2022) Financing climate-resilient coasts: tracking multilateral aid for ocean and coastal adaptation to climate change in Asia-Pacific. In: Ishiwatari M, Sasaki D (eds) Financing investment in disaster risk reduction and climate change adaptation: Asian perspectives. Springer, Tokyo
Shiohara K (2022) Factors influencing climate change adaptation investment by local government units in the Philippines. In: Ishiwatari M, Sasaki D (eds) Financing investment in disaster risk reduction and climate change adaptation: Asian perspectives. Springer, Tokyo
United Nations (2019) Report of the secretary-general on the 2019 climate action summit and the way forward in 2020. https://www.un.org/en/climatechange/assets/pdf/cas_report_11_dec.pdf
United Nations Office for Disaster Risk Reduction (UNDRR) (2021) International cooperation in disaster risk reduction. UNDRR, Geneva
United Nations Environment Programme (UNEP) (2021) Adaptation gap report 2020. UNEP, Nairobi
UNISDR (2015) Sendai framework for disaster risk reduction 2015–2030. UNISDR, Geneva
United Nations Office for Disaster Risk Reduction (UNISDR) (2005) Hyogo framework for action 2005–2015: building the resilience of nations and communities to disasters. UNISDR, Geneva
US Army Corp of Engineers (USACE) (2017) United States Army annual financial report. Department of the Army, Washington DC
World Bank (2021) Disaster risk management: overview. https://www.worldbank.org/en/topic/disasterriskmanagement/overview#2
World Meteorological Organization (2012) Associated programme on flood management. Urban flood management in a changing climate. World Meteorological Organization, Geneva

Chapter 2
Financing in Climate Change Adaptation

Ryo Fujikura

Abstract It is estimated that developing countries will need at least USD 100 billion annually for climate change adaptation by 2030. Developed countries are required to provide new and additional funds for the adaptation of developing countries and report the amount. The Organization for Economic Co-operation and Development (OECD) reports that the member countries have provided a total of USD 78.9 billion to the developing countries in 2018 for public and private financing, approaching the target set for 2020. However, there is no widely adopted definition of climate financing or no scientific consensus regarding the definition of adaptation. Developed countries have come to adopt these definitions arbitrarily. Researchers and NGOs claim that the OECD report is an overestimate and that the actual provision made by member countries was much smaller than what the developing countries actually needed. Nevertheless, multilateral development banks and internationally agreed funds such as the Global Environment Facility, Adaptation Fund, and Green Climate Fund have been providing support to facilitate the adaptation of developing countries. The private sector is also interested in investment in infrastructure, which contributes toward the enhancement of the resilience of developing countries.

Keywords Financial flows · Targets for ODA · Rio Marker · Global environment facility · Resilience bonds

2.1 Introduction

Various estimates pertaining to the scale of financing needed for the adaptation of developing countries have been presented. Chapagain et al. (2020) have estimated that the total cost for adaptation by 2030 would be between USD 50 billion and

---

R. Fujikura (✉)
Faculty of Sustainability Studies, Hosei University, Fujimi 2-17-1, Chiyodaku, Tokyo 102-8160, Japan
e-mail: fujikura@hosei.ac.jp

M. Ishiwatari and D. Sasaki (eds.), *Financing Investment in Disaster Risk Reduction and Climate Change Adaptation*, Disaster Risk Reduction,
https://doi.org/10.1007/978-981-19-2924-3_2

USD 180 billion per year, at a 50% confidence level. This will increase to USD 90–290 billion per year by 2050 in the low-risk scenario. Under the high-risk scenario, it is projected to be USD 140–450 billion per year by 2050 and USD 520 billion to USD 1.75 trillion per year by the end of the century. United Nations Environment Programme (UNEP) (2017) estimates the scale of financing required by the developing countries to be USD 280 billion to USD 500 billion per year by 2050. The World Bank also reported that the total cost of adaptation, as indicated by the Nationally Determined Contributions (NDCs) of 46 countries, would be USD 783 billion per year by 2030 (Climate Policy Initiative 2020). It is fair to say that the cost of adaptation for developing countries will be at least USD 100 billion per year by 2030.

This cost is expected to be financed primarily via public transfers coming from developed countries. The United Nations Framework Convention on Climate Change (UNFCCC) and the Paris Agreement require developed countries to provide substantial climate financing and report their amounts. However, the definition of climate financing has not been internationally adopted, and there is criticism that the amount is overstated. There are two possible reasons for this. First, while the UNFCCC, Kyoto Protocol, and Copenhagen Accord recognized that climate change assistance to developing countries should be "new and additional" financing, the ratio of Official Development Assistance (ODA) to Gross National Income (GNI) has not increased at all. This implies that no definition of climate financing has yet been adopted. Second, the definition of adaptation was not scientifically adopted. In addition, it is not possible to scientifically evaluate the increase in damages caused by climate change. As a result, it is likely that the developed countries arbitrarily adopted the definition of adaptation and overreported aid amounts. Nonetheless, the cost of damage due to natural disasters has increased. As disaster insurances become more important, the private funds looking for new sources of investment are paying more attention to it.

In this chapter, the analysis is based only on publicly available documents and data. First, the international agreements on climate change financing are reviewed. Next, the reported amount for financing climate change adaptation in developed countries is presented. Then, the claims made by researchers and NGOs regarding the overstatement of the report are presented. To investigate the reason for the overestimation, it is shown how the target of 0.7% of GNI for ODA provision has been developed. Moreover, it is shown how there has been no financing that literally corresponds to the definition of "new and additional" climate financing, which was referred to before the Paris Agreement. Then, the status of the five public international funds related to adaptation is briefly introduced, and finally, the Climate Resilience Bond, which can be regarded as a private adaptation fund, is introduced.

2.2 International Agreements and Financial Flows from Developed to Developing Countries

2.2.1 International Agreements

In terms of supporting the developing country to adapt to climate change, article 4.4 of the UNFCCC stipulates, "developed country Parties and other developed Parties included in Annex II shall also assist the developing countries that are particularly vulnerable to the adverse effects of climate change in meeting the costs of adaptation to those adverse effects." In addition, the 2007 Bali Action Plan 1, (e) (i) recognized the need to strengthen the actions on adaptation and decided to consider "improved access to adequate, predictable and sustainable financial resources and financial and technical support, and the provision of new and additional resources, including official and concessional financing for developing country Parties." The 2010 Cancun Agreements included the establishment of the Green Climate Fund to raise USD 30 billion in short-term funds from developed countries in 2012 to support the mitigation and adaptation measures carried out in developing countries from a wide range of sources (public and private, bilateral and multilateral, etc.) and raise USD 100 billion by 2020 (Kubota 2013).

In 2015, the Paris Agreement also included the provisions on adaptation finance. Article 7.1 stipulates "Parties hereby establish the global goal on adaptation of enhancing adaptive capacity, strengthening resilience and reducing vulnerability to climate change, with a view to contributing to sustainable development and ensuring adequate adaptation response in the context of the temperature goal referred to in Article 2." Article 9 requires the developed countries to "provide financial resources to assist developing country Parties with respect to both mitigation and adaptation in continuation of their existing obligations under the Convention," (Article 9.1) and "continue to take the lead in mobilizing climate finance from a wide variety of sources, instruments and channels" (Article 9.3). Provision of climate change finance shall "represent a progression beyond previous efforts," (Article 9.3) and "scaled-up financial resources should aim to achieve balance between adaptation and mitigation." (Article 9.4) Moreover, the "need for public and grant-based resources for adaptation," (Article 9.4) in particular with regard to least-developed countries and small-island developing states are recognized.

2.2.2 Financial Flows Reported by the Developed Countries

The UNFCCC guidelines (UNFCCC 2011b, Decision 2/CP.17) require Annex II parties to report about climate finance in both national communications and biennial reports and submit them to the Convention Secretariat every 4 and 2 years, respectively.

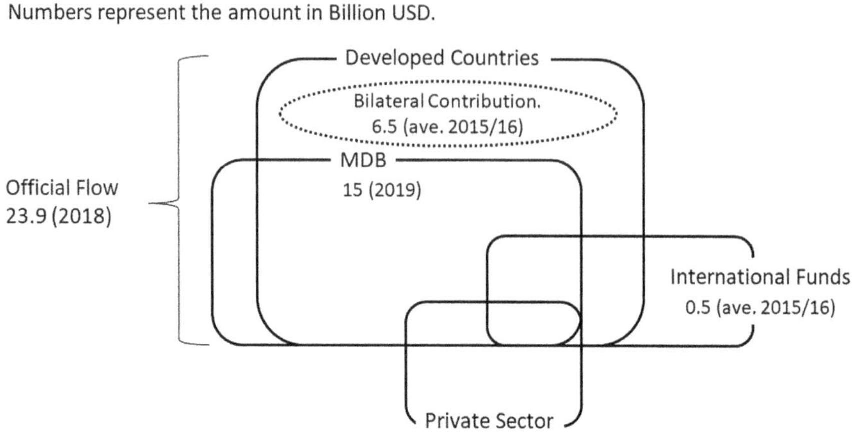

Fig. 2.1 Conceptual diagram of the flow of adaptation financing from developed countries to developing countries. *Source* OECD (2020), MDB (2020), GEF (2021a, b, c). Adaptation Fund (2021)

Figure 2.1 presents a conceptual diagram of the flow of adaptation financing from developed to developing countries. In addition to bilateral aid, developed countries provide financing through multilateral development banks and international funds. Many projects are co-financed via bilateral donors, multilateral institutions, and private funds. Thus, it is not possible to indicate the amount coming from every part shown in Fig. 2.1 based on the published data, and the size of each area does not represent the amount of financing.

The Organization for Economic Co-operation and Development (OECD) (2020) reports that the total amount of climate financing provided by developed countries to developing countries was USD 78.9 billion in 2018, an 11% increase from the USD 71.2 billion provided in 2017, while the rate of increase slowed in comparison to the 22% increase recorded in 2017 from 2016 (USD 58.6 billion). It announced that the amount is approaching the Cancun Agreements' target of USD 100 billion for climate action in 2020. Official climate financing has increased from USD 37.9 billion in 2013 to USD 62.2 billion in 2018, excluding the export credits that received climate change-related official support. If export credits are included, the figure becomes USD 64.3 billion in 2018. Private climate financing will amount to USD 14.6 billion in 2018. However, of the USD 78.9 billion, only USD 16.8 billion and USD 7.1 billion are available for adaptation and the common areas of mitigation and adaptation, respectively. The sum, i.e., $23.9 billion USD, is shown in Fig. 2.1.

The Climate Policy Initiative (CPI) (2019) estimated that both the domestic as well as international climate financing for the world in 2019 will be between USD 608 billion and USD 622 billion. In 2018, the domestic and international climate financing was USD 541 billion, with it being USD 493 billion for mitigation, USD 35 billion for adaptation, and USD 12 billion for common areas of mitigation and adaptation. Most of the mitigation amount, i.e., USD 322 billion (65% of the mitigation), was dedicated toward renewable energy and USD 115 billion (23% of the mitigation) was dedicated toward low-carbon transport. USD 388 billion (72% of the total) and USD 152 billion (28% of the total) were for domestic and foreign use, respectively. Private and public sector financing was USD 280 billion (52% of the total) and USD 261 billion (48% of the total), respectively.

The total adaptation financing was estimated to be USD 35 billion in 2018 (Climate Policy Initiative 2021). The values of the OECD and CPI are not consistent, partly because the former includes only public flow to the developing countries and the latter includes all the flows such as domestic and/or private financing. While tracking adaptation financing is difficult, as discussed below, adaptation financing has substantially increased from the 24 billion USD reported in 2017. This indicates the increasing importance of climate-resilient development and the urgency to build adaptive capacity to manage climate change vulnerabilities.

While the data for 2017/2018 are not available, as per the data for 2015/2016 (Climate Policy Initiative 2019), an annual average of USD 15 billion (69%) comes from national and multilateral development finance institutions, the primary source of adaptation finance for the developing countries. Bilateral donors contributed an annual average of USD 3.6 billion (16%) in 2015/16, and bilateral development institutions contributed USD 2.9 billion, thus, the total amount of USD 6.5 billion is shown as bilateral aid in Fig. 2.1. In addition, bilateral and multilateral climate funds contributed USD 0.5 billion.

In terms of the sectoral breakdown of adaptation, investment in disaster risk management reached USD 8 billion in 2018, second only to water and wastewater management, which stood at USD 11 billion (Fig. 2.2). The averages for 2017 and 2018 show that East Asia and the Pacific received the largest financing, accounting for 35% of the total (Fig. 2.3) (Climate Policy Initiative 2019).

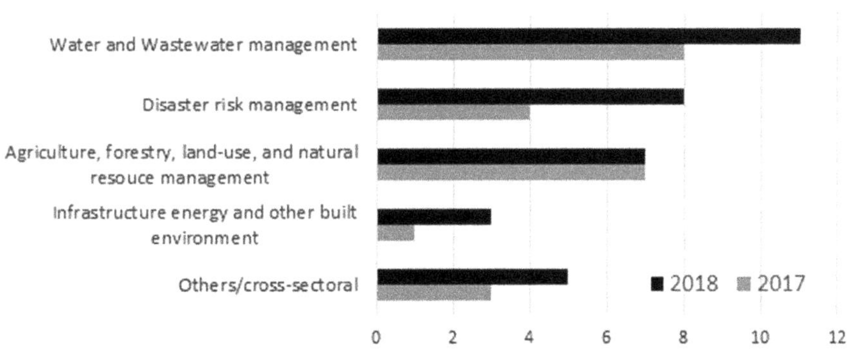

Fig. 2.2 Breakdown of global adaptation finance by sectors (USD billion) *Source* Climate Policy Initiative (2020)

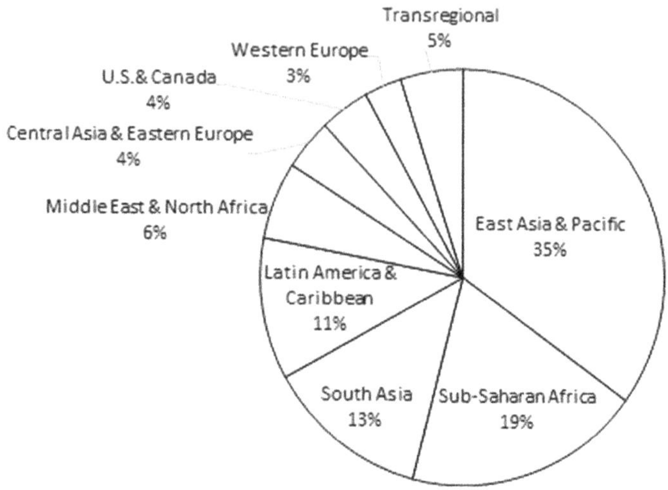

Fig. 2.3 Regional distribution of adaptation finance, 2017 *Source* Climate Policy Initiative (2020)

2.2.3 Criticism of the Amount of Assistance Reported for the Developed Countries

There has been an argument that the OECD's reported amount lacks a clear basis. Weikmans et al. (2017) reassessed the 5,200 aid activities of USD 10.1 billion reported as adaptation-related in the OECD Common Reporting Standard (CRS) database for bilateral flows in 2012. Donors reported 1,393 and 3,807 of these activities to have adaptation as the primary objective and the key objective, respectively. It also reported that nearly USD 2.7 billion of the USD 10.1 billion was for adaptation, with it being the primary objective. The reassessment concluded that 3,444 of the 5,200 aid activities, i.e., nearly USD 8 billion of the USD 10.1 billion, were not explicitly regarded as adaptation-related aid. Only USD 1.2 billion and USD 2.35 billion were deemed to be primarily dedicated for adaptation and adaptation-related activities, respectively.

Oxfam (2020) states that the OECD reported public climate financing in 2015–16 was USD 44.5 billion, increasing to USD 59.5 billion in 2017–18; however, this was a significant overestimation. Most financial flows continue to be counted at face value rather than the substantial amount given to developing countries (grant equivalent) after considering the repayment and interest. It also points out that there are problems with the way they identify climate change factors in their development projects. It argues that the amount of official net aid dedicated to climate change is much lower than reported figures, amounting to only USD 15–19.5 billion per year in 2015/16 and USD 19–22.5 billion per year in 2017/18.

2.3 Uncertainty in Financial Flows

The reason for such a large difference in the estimated financing values among organizations is that the method for estimating the amount is not only technically difficult, but it has also become a political issue. No international agreement has been reached regarding the evaluation of climate financing. In this section, it is shown that ODA targets were set without a solid basis and these targets have never been met. The definition of climate finance has not yet been clarified, and the developed countries arbitrarily define and implement climate finance. Finally, the definition of adaptation was also shown to be undefined.

2.3.1 Targets for ODA

Agenda 21, adopted by the United Nations Conference on Environment and Development (UNECD) in 1992, is an internationally agreed document that outlines what the world should do to achieve sustainable development in the 21st century. It states that "Developed countries reaffirm their commitments to reach the accepted United Nations target of 0.7% of GNP for ODA," and further states that developed countries are stipulated to make efforts "in order to reach that target as soon as possible" (Agenda 21 33.13). Since there is no significant difference between GNP and GNI, GNI will be used consistently from here on.

However, the target of 0.7% of GNI for ODA is not formulated based on solid basis, it is "an arbitrary figure based on a series of outdated assumptions going into a dubious model and measured against the wrong metric." (Clemens and Moss 2005). The debate on the need to set a target for the "flow of funds to developing countries" began at United Nation Conference on Trade and Development (UNCTAD) in the 1960s, where targets ranging from 0.6 to 1% were discussed. Considering the situation at the time and owing to a political compromise, a "simple, attainable, and adequate" target of 0.7% was adopted as an effort target by the UN General Assembly in 1970 (UNCTAD 2015). Agenda 21 reaffirmed the target.

Agenda 21 also called for "substantial new and additional funding for sustainable development" (Agenda 21 33.13). Most of the developing countries and NGOs were expecting *additional* funding from the developed countries to meet the 0.7% target. To determine if the financial flows are additional or not, a baseline must be defined, however, it is yet to be defined (Stadelmann et al. 2011).

Furthermore, in the Millennium Summit Declaration of 2000, the United Nations set the goal of doubling the world's total ODA from USD 50 billion to USD 100 billion per year by 2015. In response, the UN International Conference on Financing for Development adopted the Monterrey Consensus in 2002. However, it did not specify when the target was achieved.

In fact, after Development Assistance Committee (DAC)'s ODA/GNI ratio reached 0.54% in 1961, it continued to decline until 0.21% in 1997. The absolute

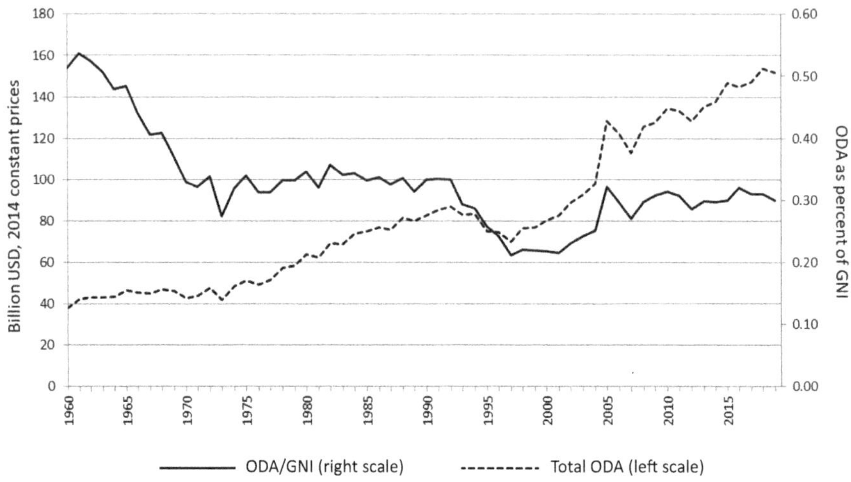

Fig. 2.4 Official development assistance over the past 50 years. *Source* OECD

amount of ODA increased from USD 134.5 billion in 2010 to USD 151.7 billion in 2019 in terms of the 2018 prices. However, as far as the GNI ratio is concerned, it is difficult to say if the new and additional funds had been provided (Fig. 2.4).

2.3.2 What is Climate Change Financing?

The phrase "new and additional" is also used in the UNFCCC, Kyoto Protocol, and Copenhagen Accord. The Paris Agreement no longer contains this phrase. Instead, Article 13 of the agreement stipulates that developed countries are obliged to report about the funds they have provided, which must be technically reviewed by experts. Even if freed from the additionality debate, there is still no agreement on what in the financial flows qualifies as adaptation finance and how it should be measured (CICERO 2017). The UNFCCC's Standing Committee on Finance has also directed the developed countries to clarify how they defined "climate-specific" (UNFCCC 2016). Even among the members of the OECD Development Assistance Committee (DAC) and the European Union (EU), there is no agreement on what constitutes climate finance (Weikmans et al. 2019).

One possible way to evaluate additionality is to calculate the incremental cost of climate change financing. For example, when it comes to mitigation, it is the additional cost to make energy efficiency higher than that required on a commercial basis. Adaptation is the additional cost of a disaster management project that assumes climate change. This may include higher seawalls than what would normally be expected. For this purpose, it is essential to set a baseline for *business as usual*, which does not assume climate change.

Since its establishment, the Global Environment Facility (GEF) has stated that the GEF shall operate for the purpose of providing new and additional grant and concessional funding to meet the agreed incremental costs for the measures needed to achieve the agreed global environmental benefits (GEF 2007). The term "business-as-usual" refers to the circumstances and context associated with a project in a given country or a proposed project site in the absence of GEF support. However, the GEF itself admits that there is a lack of understanding of the concept and procedures of incremental costs. There continues to be confusion about whether incremental costs are a form of (primarily qualitative) logic and reasoning or quantitative numerical calculations. "At a practical (and even pragmatic) level, the case for applying incremental cost assessment in a strict economic sense to GEF projects is problematic, probably unachievable, and largely unnecessary." (GEF 2007, para 5, p.2).

2.3.3 Rio Marker

The OECD/DAC compiles the climate-related ODA provided by its members each year. According to Rio Marker, countries report their aid projects under three categories. Each project is categorized as either (i) targeting the UNFCCC as a principal objective (score 2), (ii) targeting the UNFCCC as a significant objective (score 1), or (iii) not targeting the Convention (score 0) (OECD 2021).

In 2018, the OECD/DAC conducted a survey asking the member countries to voluntarily report i) if they used the Rio Marker when reporting their financial contributions to support the objectives of the UN Convention on Climate Change and Biodiversity, and ii) if so, report the coefficients they applied. It was found that the coefficients applied to the projects, which classified as "significant" (Rio Marker 1), varied from 30 to 100%, depending on the country, and that no uniform statistics were made available. According to Weikmans and Timmons Roberts (2019), six of the 24 countries applied a coefficient of 100% for the projects that were marked as significant. The United Kingdom and the United States have responded that they report to the UNFCCC in a different way than the way Rio marker does (Table 2.1).

Table 2.1 Coefficient of Rio marker "Significant" for bilateral public climate finance (2013–2014)

Coefficient	Donor
100%	Denmark, Greece, Island, Japan, Luxembourg, Norway,
50%	Austria, EU Institution, Germany, Ireland
40%	France, Italy, Netherland, Sweden
30%	Australia, New Zealand
Others	Belgium, Canada, Finland, Spain, Switzerland, UK, USA

Source Weikmans and Timmons Roberts (2019)

It is plausible that the official bilateral climate finance flows of developed countries are overestimated.

2.3.4 *What is Adaptation?*

The Intergovernmental Panel on Climate Change (IPCC) defines adaptation as "the process of adjustment to actual or expected climate and its effects." Adaptation depends on local conditions, and it requires the consideration of multiple future climate scenarios, uncertainties within those scenarios, and socioeconomic factors that contribute to vulnerability. It is almost impossible to distinguish between adaptation and "good" development. Moreover, private sector actors may refer to their activities as crisis management planning, etc., without recognizing that they are contributing to climate change adaptation.

To identify whether a development project constitutes an adaptation, it is necessary to understand what will happen in the context of climate risk taking cultural, political, socioeconomic, and environmental factors into consideration. However, it is very difficult to conduct such a detailed analysis of every one of the many development projects and calculate the adaptation cost portion (Hammill and McGray 2018). There are still no widely accepted standards or benchmarks to measure adaptation (Uitto 2021), and the Global Center on Adaptation (2020) proposes that a core set of indicators be developed and agreed upon, given the difficulties in tracking adaptation financing due to insufficient data, difficulties in establishing baselines, and the lack of a single indicator or a small set of indicators to monitor adaptation.

Japan International Cooperation Agency (JICA), for example, made a distinction between "adaptation projects" and "conventional development plus adaptation options" in the pilot version of the Climate Finance Impact Tool for Adaptation, whose final version was officially adopted in 2019. "Adaptation" projects are explicitly designed to reduce the vulnerability of the current system to anticipated climate change. For example, it is a response to the emergence of vulnerability in the case of external force changes due to climate change in a system where facilities are already in place. "Conventional development plus Adaptation Option" is a project where the main objective is not to address the reduction of vulnerability but takes this vulnerability into account. For example, a project that considers post-climate change external forces, instead of current external forces, in planning and designing a new or rehabilitated infrastructure (JICA 2011). Since the Japanese government uses the Rio Marker, "adaptation projects" seem to be classified as principal objectives (score 2), and "conventional development + adaptation options" seem to be classified as significant objectives (score 1). In both cases, the entire amount of the project cost would be accounted for as adaptation financing, regardless of the extent to which the adaptation elements are incorporated.

The Multilateral Development Banks (MDBs) adopted the following in 2015 as a step to track adaptation finance without adopting the Rio marker (MDB 2020):

Table 2.2 MDB adaptation finance in accordance to the source of funds, 2019 (in USD million)

MDB	MDB own account	MDB-managed external resources	Total
African Development Bank	1,695	321	2,016
Asian Development Bank	1,413	123	1,536
European Bank for Reconstruction and Development	567	14	582
European Investment Bank	936	35	971
Inter-American Development Bank Group	1,824	94	1,918
Islamic Development Bank	218	–	218
World Bank Group	7,336	361	7,697
Total	13,989	948	14,937

Source MDB (2020)

1. Setting out the context of risks, vulnerabilities, and impacts related to climate variability and climate change.
2. Stating the intent to address the identified risks, vulnerabilities, and impacts on project documentation.
3. Demonstrating a direct link between the identified risks, vulnerabilities, impacts, and financed activities.

In this way, only the specific projects that clearly address climate change vulnerability are regarded as adaptation. Then, the incremental costs are calculated and considered as adaptation finance.

In 2019, the MDBs reported a total commitment of USD 14,937 million for climate change adaptation financing (Table 2.2). According to the report, this financing corresponds to the incremental cost of a component, sub-component, element of a project, or a proportion of a project, which aims to reduce the vulnerability to climate change and build resilience to climate change.

2.4 International Funds Related to Adaptation

There are international funds that actively provide climate finance, and some of them are exclusively meant for adaptation projects, although there is no international consensus on climate finance and adaptation.

2.4.1 Global Environment Facility (GEF) and the Funds Related to It for Adaptation

The GEF is a trust fund that was established at the World Bank as a financial mechanism for five environment-related conventions, including the UNFCCC. International organizations such as the World Bank, United Nations Development Programme, and UNEP implement projects using GEF funds. It began with a pilot phase launched in May 1991 for a period of three years based on the resolution of the World Bank Board of Governors. In 1992, the GEF was formally established as an operating entity of the financial mechanism of the UNFCCC to provide financing for country-driven climate change mitigation and climate change adaptation projects. Based on the knowledge and expertise accumulated by the end of the pilot phase, the GEF was officially launched in 1994. Since then, the GEF has been replenished every four years and is currently in the GEF-7 period.

The GEF provides *incremental* costs to developing countries to implement the projects that address global environmental issues. Funds are not provided to meet the entire cost of individual projects. They are provided to meet the incremental costs only. This has the significance of providing leverage to ensure that development projects are formulated in an environmentally friendly manner. In exceptional cases, 100% of the funds are provided for projects that are essential to realize the objectives of the conventions, such as the preparation of national country reports under the UNFCCC. As of January 2021, USD 1.9 billion had been provided in grant financing for 386 adaptation projects in 98 countries, reducing the vulnerability of almost 27 million people in more than 130 countries (GEF 2021a).

The challenge for the GEF right from the beginning was that it considered adaptation to be a *national* development issue, while it has explicitly concerned itself with *global* environmental benefits. Therefore, adaptation has never been formally qualified for GEF financing. The GEF then finances the Least Developed Countries Fund (LDCF) and the Special Climate Change Fund (SCCF), which were designated to serve the Paris Agreement along with the Green Climate Fund (GCF).

The LDCF was established in response to the guidance from COP7 in Marrakesh in 2001. It aims to finance the special needs of the least developed countries under the convention with the priority to prepare and implement the National Adaptation Programs of Action (NAPAs). As of 2021, it has financed 305 initiatives by providing USD 1.6 billion in grants (GEF 2021b).

The SCCF was also established at COP7 to support adaptation and technology transfer in *all* the developing country parties of the UNFCCC. As of 2021, it has invested USD 355 million in 87 projects, benefiting 7 million people around the world and supporting 7,500 assessments of risk, vulnerability, and other factors (GEF 2021c).

2.4.2 Adaptation Fund (AF)

The Adaptation Fund was established as a fund administered by the GEF under the Marrakech Accord, adopted at COP7 in 2001. It is a fund under the Kyoto Protocol and it receives funding from 2% of the Certified Emission Reductions (CERs) issued under the Protocol's Clean Development Mechanism projects. Other major sources of financing are governments and private donors. It is unique in the sense that the majority of board members are from developing countries. It has adopted the direct access modality, which allocates project funds directly to the national implementing entities approved by the board. This approach was introduced to encourage self-help and increase ownership by allowing developing countries to plan, implement, and manage their own projects. As of June 30, 2020, the AF has approved USD 740 million in grants for projects and programs, including 107 specific projects (Adaptation Fund 2020).

2.4.3 Green Climate Fund (GCF)

The GCF was established in 2010 as an international fund to support developing countries in their efforts to implement climate change measures. It provides co-financing of public and private funds to create a "paradigm shift" in development, namely low emission and climate resilience. It ensures the project ownership of developing countries by encouraging them to propose and manage projects in line with their development plans and climate change policy priorities.

The funds are allocated half to mitigation and half to adaptation. In consideration of the regional balance, at least 50% of the amount allocated for adaptation will be allocated to the poorest countries (LDCS), small island developing states (SIDS), and African countries. From 2015 to 2020, 159 projects were approved, amounting to a total of USD 23 billion. Of this amount, the GCF allocation was USD 7.3 billion, with USD 4.6 billion for mitigation and USD 2.7 billion for adaptation. In grant equivalent terms, mitigation and adaptation are 50% of the total, as originally stipulated (GCF 2021).

2.5 Private Financing

While the public financial flows for adaptation are increasing, uncertainty remains regarding the reported amounts. Nevertheless, the financing gap will widen as the adaptation financing needed by developing countries increases rapidly in the future under any scenario. One possibility of new adaptation finance flow from developed countries to developing countries has been proposed as debt for climate swaps. This mechanism is similar to Debt for Nature Swap, in which the debtor countries invest in climate projects and creditors forgive a part of the debt (Global Center on Adaptation 2020).

The ratio of ODA from developed countries to GNI has stagnated, and there is no expectation of a sharp increase in public financing from developed countries. Even if new channels of public finance such as Debt for Nature Swap are opened for adaptation, they will not be large enough to address this gap. If public financing for climate change, including adaptation, does expand significantly under such a situation, it could crowd out the support provided for other purposes. Rather, attention should be paid to the private sector financial flows related to adaptation. Disaster risk-sharing insurance can be seen as a form of adaptation that can be financed through the bond market. This insurance has come to attract considerable attention. It could contribute to facilitating adaptation in developing countries in the future.

2.5.1 Catastrophe Bonds (CAT Bonds)

While insurance risks are diversified by reinsurance, the capacity of reinsurance is not sufficient to increase the risks. However, the capital markets are large. While the reinsurance market is worth USD 3.3 trillion, the total global bond market is worth USD 45 trillion. Bond markets can be used to diversify huge catastrophe risks.

Catastrophe bonds (CAT bonds) are structured bonds that pay higher interest rates than the straight bonds issued by issuers with similar ratings. Instead, the redemption principal for investors will be reduced in the event of natural disasters such as typhoons, floods, and earthquakes, in accordance with the conditions set in the contract in advance. For the issuing company, there are no restrictions on how the funds received can be used, allowing for a wide range of responses to disaster recovery and the ability to receive funds in a relatively short period of time after the disaster. In recent years, the demand for CAT bonds has increased among the institutional investors in Europe and the United States, as it is a risk diversification product that has a low correlation with the economy, stock prices, and interest rate fluctuations, and can provide high returns. The market size has been expanding year by year, with the issuance amount reaching USD 16.4 billion in 2020 and the outstanding amount reaching USD 46.4 billion (Artemis 2021).

2.5.2 Resilience Bonds

The resilience bond is an elaboration of the CAT bond. It is designed to be used for infrastructure development, such as disaster management. The funds raised through the bond will be specifically earmarked for climate change resilience projects, such as building seawalls to tackle the rising sea levels and building shelters in areas vulnerable to cyclones and typhoons. If the government builds disaster-resilient infrastructure, the risk held by the issuer of the CAT bond will be reduced, which in turn will reduce the premium. Premium reduction is then returned to the government. In many cases, when the bond matures, the investor will not only receive the principal but also the interest accrued during the life of the bond, resulting in a better return on investment.

The first resilience bonds are the climate resilience bonds (CRBs) issued by the European Bank of Reconstruction and Development (EBRD) in 2019. There will be more opportunities available to the financing projects to enhance resilience, as EBRD will reduce the vulnerability and risks of climate change identified by the public and private sector projects supported by CRBs. CRB proceeds will be used to support the Climate Resilience Portfolio, which consists of investments in climate-resilient infrastructure, business and commercial activities, agriculture, and ecosystems. The five-year bond attracted demand from 40 investors across 15 countries. As of December 31, 2020, the CRBs had EUR 975 million outstanding issues, with 74 projects falling under management (EBRD 2021).

2.6 Discussion

To demonstrate the extent to which an adaptation component is incorporated into a disaster management project, a baseline must be established scientifically, assuming no climate change. This is impossible because economic "growth, including greater concentrations of people and wealth in periled areas and rising insurance penetration, is the most important driver of increasing losses," and the "loss trends have not been conclusively attributed to anthropogenic climate change" (IPCC 2014). Pielke (2021) reviewed 54 normalized studies published between 1998 and 2020, and found that global economic losses were recorded on a climate time scale. He found that there is little evidence to support the claim that the overall increase in global economic losses is due to anthropogenic climate change, reinforcing the findings of the IPCC assessment.

While the IPCC states that adaptation and disaster risk management need to be integrated (IPCC 2014), it is also the case that the two have been addressed in different international processes. In the Sendai Framework for Disaster Risk Reduction 2015–2030, for example, the footnote states the following: "The climate change issues mentioned in this Framework remain within the mandate of the United Nations Framework Convention on Climate Change under the competences of the Parties to the Convention." There was no clear indication that disaster risk management should be actively promoted as an adaptive measure.

Whether a project is adopted is determined politically, not scientifically. However, it is no longer politically feasible to regard a project that reduces disaster risks as a conventional disaster management project without any element of adaptation. Therefore, to secure financing for disaster reduction projects, it is important for donors to provide detailed, evidence-based descriptions of climate change impacts and associated vulnerabilities, and clearly argue that they are adaptation-related. It is hoped that the accumulation of scientific knowledge will make it possible to estimate the incremental cost of disaster risk management related to climate change. Moreover, the growing interest in CAT bonds and CRBs will open the way to attract more private funds for disaster management projects.

Acknowledgements This study was partly supported by the Collaborative Research Project of the International Research Institute of Disaster Science, Tohoku University. The author would like to thank several MDB and GEF officials for their helpful comments and suggestions.

References

Adaptation Fund (2021) Annual Performance Report 2020. https://www.adaptation-fund.org/document/annual-performance-report-for-fiscal-year-2020/. Accessed 23 Jan 2021
Artemis (2021) Catastrophe bonds & ILS issued and outstanding by year. https://www.artemis.bm/dashboard/catastrophe-bonds-ils-issued-and-outstanding-by-year/. Accessed 20 July 2021
CICERO (2017) Mobilizing adaptation finance in developing countries. Report 2017:2. https://pub.cicero.oslo.no/cicero-xmlui/bitstream/handle/11250/2435614/Torvanger%202017%2002%20web.pdf?sequence=1&isAllowed=y. Accessed 25 Mar 2021
Clemens MA, Moss TJ (2015) Ghost of 0.7%: origins and relevance of the international aid target. Working Paper Number 68, September 2005, Center for global Development
Climate Policy Initiative (2019) Global landscape of climate finance. https://www.climatepolicyinitiative.org/publication/global-landscape-of-climate-finance-2019/. Accessed 25 Mar 2021
Climate Policy Initiative (2020) A snapshot of global adaptation investment and tracking methods. https://climatepolicyinitiative.org/wp-content/uploads/2020/04/A-Snapshot-of-Global-Adaptation-Investment-and-Tracking-Methods-April-2020.pdf. Accessed 25 Mar 2021
Climate Policy Initiative (2021) Updated view on the global landscape of climate finance 2019. https://www.climatepolicyinitiative.org/publication/updated-view-on-the-global-landscape-of-climate-finance-2019/. Accessed 25 Mar 2021
Conventions on Climate Change and Biodiversity (2019). OECD-DAC
D Chapagain F Baarsch M Schaeffer 2020 Climate change adaptation costs in developing countries: insights from existing estimates Climate Dev 12 10 934 942 https://doi.org/10.1080/17565529.2020.1711698
EBRD (2021) Focus on environment. https://www.ebrd.com/documents/treasury/ebrd-focus-on-environment.pdf?blobnocache=true. Accessed 20 June 2021
GEF (2007) Operational guidelines for the application of the incremental cost principle. In: GEF/C.31/12
OECD (2021) OECD DAC Rio markers for climate handbook. https://www.oecd.org/dac/environment
GEF (2021a) Financing adaptation to climate change. https://www.thegef.org/sites/default/files/publications/GEF_Adaptation%20Bifold_2021a_r2.pdf. Accessed 1 June 2021a
GEF (2021b) Least developed countries fund—LDCF. https://www.thegef.org/topics/least-developed-countries-fund-ldcf. Accessed 20 July 2021b
GEF (2021c) Special climate change fund—SCCF. https://www.thegef.org/topics/special-climate-change-fund-sccf. Accessed 20 July 2021c
Global Center on Adaptation (2020) States and trends in adaptation report 2020. https://gca.org/reports/state-and-trends-in-adaptation-report-2020/. Accessed 22 July 2021
Green Climate Fund (2021) Annual results report. https://www.greenclimate.fund/annual-results-report-2020. Accessed 15 Jun 2021

Hammill A, McGray H (2018) Is it adaptation or development? Revisiting the continuum 10 years later March 5, 2018. https://www.iisd.org/story/is-it-adaptation-or-development/. Accessed 11 June 2021

Ikeda M (2018) 開発途上国における気候変動適応とモニタリング・評価 (in Japanese). Dissertation, Kyoto University. https://doi.org/10.14989/doctor.k20822

IPCC (2014) Climate change 2014: impacts, adaptation, and vulnerability. Part A: global and sectoral aspects. In: Contribution of working group II to the fifth assessment report of the intergovernmental panel for climate change. Cambridge University Press, pp 659–708.

JICA (2011) Climate finance impact tool for adaptation (provisional version) (in Japanese). https://openjicareport.jica.go.jp/pdf/12047932.pdf. Accessed 23 Jan 2021

Kubota I (2013) A qualitative comparative analysis of international financial mechanisms for adaptation to climate change impact (in Japanese). 環境情報科学 学術研究論文集 27:243–246

MDB (2020) The joint report on multilateral development banks' climate finance. https://www.eib.org/attachments/press/1257-joint-report-on-mdbs-climate-finance-2019.pdf. Accessed 3 Mar 2021

OECD (2020) Climate finance provided and mobilised by developed countries in 2013–18. OECD Publishing, Paris. https://doi.org/10.1787/f0773d55-en

OECD DAC (2021) Rio markers for climate handbook. https://www.oecd.org/dac/environment-development/Revised%20climate%20marker%20handbook_FINAL.pdf. Accessed 28 Jan 2021

Results of the first survey on coefficients that members apply to the Rio marker data when reporting to the UN

Oxfam (2020) Climate finance shadow report 2020. https://www.oxfam.org/en/research/climate-finance-shadow-report-2020. Accessed 2 Feb 2021

R Pielke 2021 Economic 'normalisation' of disaster losses 1998–2020: a literature review and assessment Environ Hazard 20 2 93 111 https://doi.org/10.1080/17477891.2020.1800440

M Stadelmann JT Robert A Michaelowa 2011 New and additional to what? assessing options for baselines to assess climate finance pledges Climate Dev 3 175 192 https://doi.org/10.1080/17565529.2011.599550

UNCTAD (2015) New and additional climate finance: a continuing lack of clarity. Policy Brief No.41, 2015. https://unctad.org/news/new-and-additional-climate-finance-continuing-lack-clarity. Accessed 28 Jan 2021

UNEP (2017) The adaptation gap report 2016. https://climateanalytics.org/publications/2016/the-adaptation-gap-report-2016/. Accessed 19 Feb 2021

UNFCCC (2016) Report of the standing committee on finance to the conference of the parties. FCCC/CP/2016/8. https://unfccc.int/documents/184363. Accessed 11 June 2021

Uitto JI (2021) Evaluation at the nexus between nature and humanity for transformational change. In: van den Berg RD, Magro C, and Adrien M (eds) Transformational evaluation for the global crises of our times. International Development Evaluation Association (IDEAS), Exeter, UK, pp 87–103

Weikmans R, Roberts JT, Baum J et al (2017) Assessing the credibility of how climate adaptation aid projects are categorized. Dev Pract 27(4):458–471. https://doi.org/10.1080/09614524.2017.1307325

https://www.cgdev.org/sites/default/files/3822_file_WP68.pdf. Accessed 17 Mar 2021

R Weikmans JT Timmons Roberts 2019 The international climate finance accounting muddle: is there hope on the horizon? Climate Dev 11 2 97 111 https://doi.org/10.1080/17565529.2017.1410087

development/Revised%20climate%20marker%20handbook_FINAL.pdf. Accessed 28 Jan 2021

京都大学防災研究所 (2015) 大規模都市災害からの経済回復 (in Japanese). http://www.drs.dpri.kyoto-u.ac.jp/ur/reports/h27/pdf/3.2.1.3.pdf. Accessed 15 Jun 2021

Chapter 3
Recent Trends in Disaster Risk Reduction Investments: A Literature Review

Daisuke Sasaki and Mikio Ishiwatari

Abstract In this study, we reviewed the literature on disaster risk reduction investment published since 2015, when the Sendai Framework for Disaster Risk Reduction was adopted, and then followed up with text mining analysis. The literature extracted from the databases of the World Bank and the Asian Development Bank (ADB), which are multilateral development banks (MDBs), offer new concepts in disaster risk reduction investment, such as the triple dividend of resilience and women-focused investment in resilience. Additionally, we found that solutions such as the Blueprint for Action (World Bank) had been proposed to promote investment. Conversely, many of the literature extracted from the Web of Science, a database of academic papers, are based on empirical research using quantitative methods and can be regarded as providing a certain level of evidence for the concepts that have been discussed in MDBs. The results of the co-occurrence network show that words such as "agriculture" and "fund" are characteristic of the World Bank literature, while "woman" and "community" are characteristic of the ADB literature. Moreover, the correspondence analysis implies a difference between the tendency of words to appear in the Web of Science literature and that in other MDB databases. Thus, when contemplating the direction of future academic research, the characteristic words of literature extracted from the MDB database using text mining could be deemed useful as a source of reference information to a certain extent.

Keywords Disaster risk reduction investments · Sendai Framework for Disaster Risk Reduction 2015–2030 (SFDRR) · Multilateral development banks (MDBs) · Literature review · Text mining analysis

---

D. Sasaki (✉)
International Research Institute of Disaster Science (IRIDeS), Tohoku University, Sendai, Miyagi, Japan
e-mail: dsasaki@irides.tohoku.ac.jp

M. Ishiwatari
Graduate School of Frontier Sciences, The University of Tokyo, Chiba, Japan
e-mail: ishiwatari1219@gmail.com

M. Ishiwatari and D. Sasaki (eds.), *Financing Investment in Disaster Risk Reduction and Climate Change Adaptation*, Disaster Risk Reduction,
https://doi.org/10.1007/978-981-19-2924-3_3

3.1 Introduction

The Sendai Framework for Disaster Risk Reduction 2015–2030 (SFDRR), adopted at the Third UN World Conference on Disaster Risk Reduction (UNWCDRR), held in March 2015, in Sendai, Japan, presents the four priorities for action, namely "Understanding disaster risk"; "Strengthening disaster risk governance to manage disaster risk"; "Investing in disaster risk reduction for resilience"; and "Enhancing disaster preparedness for effective response" and to "Build Back Better" in recovery, rehabilitation, and reconstruction (UNDRR 2015). It is of great significance that investment in disaster risk reduction (DRR) has been emphasized in a global DRR agenda, such as the SFDRR. This implies that it is widely recognized that public and private investment in disaster risk prevention and reduction through both structural (hard measures) and non-structural (soft measures) measures is essential, as these measures are cost-effective and beneficial in saving lives, preventing or reducing economic losses, and achieving effective recovery and reconstruction.

The outcome of the Third UNWCDRR was reflected in the 2030 Agenda for Sustainable Development, which was adopted at the UN Sustainable Development Summit in September 2015, in which we can find the perspectives on DRR in the form of its interrelatedness with the SFDRR (UN 2015); for instance, Target 11.b of Goal 11 "Make cities and human settlements inclusive, safe, resilient, and sustainable" denotes "By 2020, substantially increase the number of cities and human settlements adopting and implementing integrated policies and plans toward inclusion, resource efficiency, mitigation and adaptation to climate change, and resilience to disasters and develop and implement, in line with the Sendai Framework for Disaster Risk Reduction 2015–2030, holistic disaster risk management at all levels." Target 11.5 of Goal 11 requires that "By 2030, significantly reduce the number of deaths and the number of people affected and substantially decrease the direct economic losses relative to the global gross domestic product caused by disasters, including water-related disasters, with a focus on protecting the poor and people in vulnerable situations." Furthermore, Target 1.5 of Goal 1 "End poverty in all its forms everywhere" asserts that "By 2030, build the resilience of the poor and those in vulnerable situations and reduce their exposure and vulnerability to climate-related extreme events and other economic, social, and environmental shocks and disasters." Also, Target 13.1 of Goal 13 "Take urgent action to combat climate change and its impacts" denotes "Strengthen resilience and adaptive capacity to climate-related hazards and natural disasters in all countries." Thus, the SFDRR and the Sustainable Development Goals (SDGs) are interrelated, and the significance of investment in DRR can be viewed in the context of the SDGs.

In December 2015, the Conference of the Parties to the United Nations Framework Convention on Climate Change (UNFCCC) adopted the Paris Agreement, which entered into force in 2016. The Paris Agreement is a new international framework for reducing greenhouse gas emissions after 2020, replacing the Kyoto Protocol. Article 2 defines its purpose as "Making finance flows consistent with a pathway toward low greenhouse gas emissions and climate-resilient development" (UNFCCC

2016). Furthermore, Article 9 states "Developed country Parties shall provide financial resources to assist developing country Parties with respect to both mitigation and adaptation in continuation of their existing obligations under the Convention" (Section 1); "As part of a global effort, developed country Parties should continue to take the lead in mobilizing climate finance from a wide variety of sources, instruments and channels, noting the significant role of public funds, through a variety of actions, including supporting country-driven strategies and taking into account the needs and priorities of developing country Parties. Such mobilization of climate finance should represent a progression beyond previous efforts" (Section 3) (UNFCCC 2016). Thus, the importance of investment is increasing in the context of climate resilience.

Additionally, initiatives related to sustainable finance have accelerated recently; for instance, in the EU, a legally binding taxonomy regulation came into force in July 2020 (EU 2020). This regulation establishes six environmental objectives in Article 9: climate change mitigation, climate change adaptation, sustainable use and protection of water and marine resources, transition to a circular economy, pollution prevention and control, and protection and restoration of biodiversity and ecosystems. The European Commission will adopt delegated acts to define technical screening criteria for each environmental objective, thereby creating a list of environmentally sustainable activities. The background to this taxonomy is considered to be the SDGs and the Paris Agreement, and these efforts are expected to accelerate in the future.

In the area of international standardization of sustainable finance, the International Organization for Standardization (ISO) has been active in recent years. Table 3.1 shows recent ISO trends in environmental finance. As you can see, ISO14097 (Greenhouse gas management and related activities—Framework including principles and requirements for assessing and reporting investments and financing activities related to climate change), an international standard (IS) proposed by France, was published in May 2021 (ISO 2021a). Other Final Draft International Standards (FDISs) and Committee Drafts (CDs) have been proposed by several countries and are currently in the process of being developed for publication, including FDISs that specify the process for green bonds and loans. In light of this situation, ISO international standards are expected to play a role in the future in stimulating investment in climate change adaptation. The development of international standards by ISO concerning DRR is currently underway (UNDRR 2021). Risk finance, which is closely related to investments in DRR, also falls within this scope. It is expected that new international standards by the ISO will be published soon to further promote investments in DRR.

Table 3.2 summarizes the items in the global agendas and ISs of the ISO that are relevant to DRR investments. These items are interrelated and closely linked with each other; thus, the authors attempt to grasp investments in DRR more broadly and holistically, such as by including climate change adaptation in this study. The objective of this study is to review the literature on investments in DRR published since 2015, when the SFDRR was adopted, to summarize the current status of investments in DRR, and to identify distinctive terms and differences in the literature through coherent text mining with the narrative results of the literature review

Table 3.1 Recent ISO trends in environmental finance (as of July 2021)

Title	Technical committee	History
ISO 14097:2021 Greenhouse gas management and related activities—Framework including principles and requirements for assessing and reporting investments and financing activities related to climate change	ISO/TC207/SC7 Greenhouse gas management and related activities	New project approved in February 2017 International Standard published in May 2021
ISO 14030 Environmental performance evaluation—Green debt instruments Part 1: Process for green bonds Part 2: Process for green loans Part 3: Taxonomy Part 4: Verification program requirements	ISO/TC207/SC4 Environmental performance evaluation	New project approved between July 2017 and November 2018 Under development [FDIS (Parts 1, 2, and 4); DIS (Part 3)]
ISO 14100 Green Finance: Assessment of Green Financial Projects	ISO/TC207/SC4 Environmental performance evaluation	New project approved in June 2018 Under development [CD]
ISO 32210 Framework for sustainable finance: Principles and guidance	ISO/TC322 Sustainable finance	New project approved in October 2019 Under development [CD]

Source ISO (2021a, b, c, d, e, f, g)

Table 3.2 Summary of the items relevant to DRR investments

SFDRR	SDGs	Paris Agreement	ISs of the ISO
• Priority 3	• Goal 1 (Target 1.5) • Goal 11 (Target 11.5, Target 11.b) • Goal 13 (Target 13.1)	• Article 2.1(c) • Article 9	• ISO 14097:2021 • ISO 14030 (*) • ISO 14100 (*) • ISO 32210 (*) (*) under development

The results of this study were initially presented at the 22nd Spring Conference of the Japan Society for International Development in Japanese and later expanded into this chapter for a more general readership (Sasaki 2021).

3.2 Methodology

In this study, we used three databases (Open Knowledge Repository [the World Bank], Publications and Documents [the Asian Development Bank (ADB)], and Web of Science) to identify literature on DRR investment published since 2015, when the SFDRR was adopted, that had the word "invest" in addition to the words "disaster" or "resilience" in their titles. As for the Publications and Documents by the ADB, the authors selected literature whose subject was "Disaster Risk Management" and which included the word "invest" in their titles. However, literature that was not related to DRR investments was excluded. The reason we used the databases of multilateral development banks (MDBs), namely the World Bank and the ADB, is that this study focused on the recent trends of practical investments in DRR in Asia. The literature stored in the databases of these two MDBs is supposed to adequately reflect such trends.

KH Coder is an open-source, free software that extracts words from the English language data using Stanford POS Tagger, performs statistical analysis using R, and organizes and searches data using MySQL (Higuchi 2016, 2017). The literature to be analyzed was converted into text data after tagging the database and then combined to form a corpus, a set of text documents to be analyzed.

3.3 Results and Discussion

3.3.1 Narrative Summary of the Literature Review

Based on the extraction described in Sect. 3.2, the authors have identified 24 works including seven from the World Bank's Open Knowledge Repository, nine from Publications and Documents by the ADB, and eight from the Web of Science. Considering the limitation that only literature on DRR investment published since 2015 was targeted, the authors deem the number of the extracted literature to be acceptable.

In the literature extracted from the World Bank's Open Knowledge Repository, there are some references on the importance of private investment in DRR. Tall et al. (2021), for example, argue that public finance alone cannot provide the funds needed for climate change adaptation and resilience and that it is essential to fill the funding gap by leveraging private investment. As shown in Fig. 3.1, the Overseas Development Institute (ODI) and the World Bank (2015) proposed the concept of "Triple Dividend of Resilience." The first dividend of resilience is the avoided losses in the event of a disaster striking. The ODI and the World Bank (2015) advocate that by increasing resilience, it is possible to reduce disaster losses by saving lives and mitigating infrastructure damages and economic losses. This can be regarded as a conventional benefit concept for disaster risk management projects. Meanwhile, the second and third dividends of resilience are "Development" and "Co-benefits," respectively. Development refers to the release of economic potential and stimulation

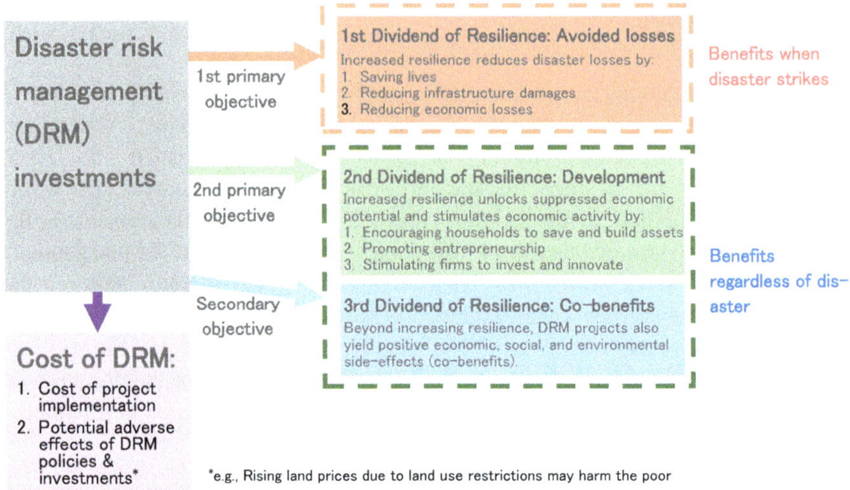

Fig. 3.1 The triple dividend of resilience. *Source* Overseas Development Institute & World Bank (2015)

of economic activities by encouraging households to save and build assets, promoting entrepreneurship, and stimulating firms to invest and innovate; co-benefits refer to the economic, social, and environmental benefits resulting from the implementation of disaster risk management projects. Unlike the first dividend of resilience, "Avoided losses," the second and third dividends of resilience can be said to be unique in that these benefits accrue regardless of whether or not a disaster occurs. Besides, private investment will also enhance the profitability of the private sector.

However, it has long been pointed out that there are significant obstacles to the introduction of private investment; for instance, the World Bank (2016) points out that the introduction of private investment in building urban resilience is challenged by a lack of experience and capacity of the developing country governments as well as their lack of creditworthiness. According to Tall et al. (2021), the World Bank has developed a "Blueprint for Action" to accelerate private investment in climate change adaptation and resilience building, which will be tested in five pilot countries from 2021 to 2023. The Blueprint for Action recommends five steps: first, public sector-led support for long-term adaptation planning; second, public sector-led development of national adaptation investment plans; third, pipeline screening and market assessment through public–private partnerships; fourth, project preparation support through public–private partnerships; and fifth, downstream transaction demonstration led by private sector initiative. In this way, the Blueprint for Action is divided into three areas: upstream, midstream, and downstream, and each step is set up accordingly.

In the literature extracted from Publications and Documents by the ADB, there exists literature that refers to the "Triple Dividend of Resilience" concept. For instance, Mechler and Hochrainer-Stigler (2019) pointed out that there have been a few studies that have conducted impact assessments on triple dividends and attempted to present evidence on the dividend of resilience. Moreover, some concepts such as "women-focused investment" and community-based flood risk management in investment have been proposed (ADB 2020; Osti 2017). According to the ADB (2020), women-focused investments are supposed to simultaneously realize gender equality as well as climate and disaster resilience building, both of which are two of the seven operational priorities of the ADB's Strategy 2030. Women-focused investment is a concept of investing in projects led by women in various sectors, both in the public and private sectors. Furthermore, the ADB (2016) points to the underestimation of risks of less-frequent hazards as well as weak political support due to the lack of immediate results.

Many of the literature extracted from the Web of Science, a database of academic papers, are based on empirical research using quantitative methods, unlike the literature from the databases of the World Bank and the ADB, which are MDBs. For example, Fung et al. (2020) attempted to quantify the dividends of resilience using computable general equilibrium modeling. Additionally, Banerjee et al. (2018) conducted a case study to estimate the benefits derived from investments in DRR in Small Island Developing States using the contingent valuation method. Moreover, Young et al. (2017), using experimental economics, analyzed the impact of biological sex or gender diversity on decision-making regarding investments in resilience. Thus, we can see that some recent literature based on academic research supports the concept, such as the Triple Dividend of Resilience, discussed by the World Bank and the ADB.

3.3.2 Text Mining Analysis

In text mining, the literature to be analyzed is tagged with the database name from which it was extracted (the World Bank [Open Knowledge Repository], the ADB [Publications and Documents], and Web of Science) and then converted into text data. As a result, the word "climate" has been output as a characteristic word in the documents extracted from the World Bank and the ADB databases. The co-occurrence network shown in Fig. 3.2 also shows that "climate" (drawn in the upper left corner) has a co-occurrence relationship with both the World Bank literature and the ADB literature. This implies that investments in DRR and climate change adaptation tend to be discussed in an integrated manner in the literature of the World Bank and the ADB. Moreover, the co-occurrence network visualizes common words among the literature retrieved from each database. In this context, it is found that words such as "investment," "risk," "country," "resilience," and "disaster" commonly appear in every literature. The results of the co-occurrence network in Fig. 3.2 show that words such as "agriculture" and "fund" are characteristic of the World Bank literature, while words such as "woman" and "community" are characteristic of the ADB literature.

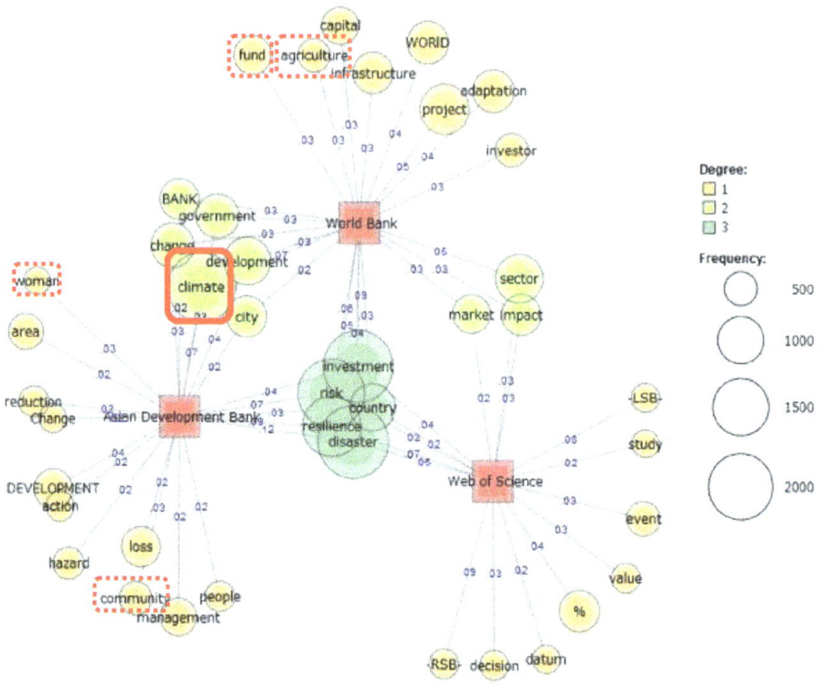

Fig. 3.2 The co-occurrence network. *Source* Sasaki (2021)

The results of the correspondence analysis are presented in Fig. 3.3. In the correspondence analysis, tags of the database and words are simultaneously plotted on the same plane; the farther away the words are plotted from the origin, the more we interpret them as word characteristics of tags, namely the name of the databases, in that direction. As opposed to the co-occurrence network, the correspondence analysis shows the characteristic words and the tendency of words to appear in each database more visibly. Consequently, we can see that words such as "agriculture" and "fund" are characteristic of the World Bank literature, and words such as "woman" and "community" are characteristic of the ADB literature. Additionally, it is suggested that there is a difference in the tendency of words to appear in the Web of Science literature and that of other MDB databases. As already mentioned, academic papers are recognized to provide a certain level of evidence for the concepts discussed in the MDBs. Therefore, it is also considered to be effective to a certain extent to refer to the characteristic words of the literature extracted from the MDB databases when considering the possible direction of future academic research.

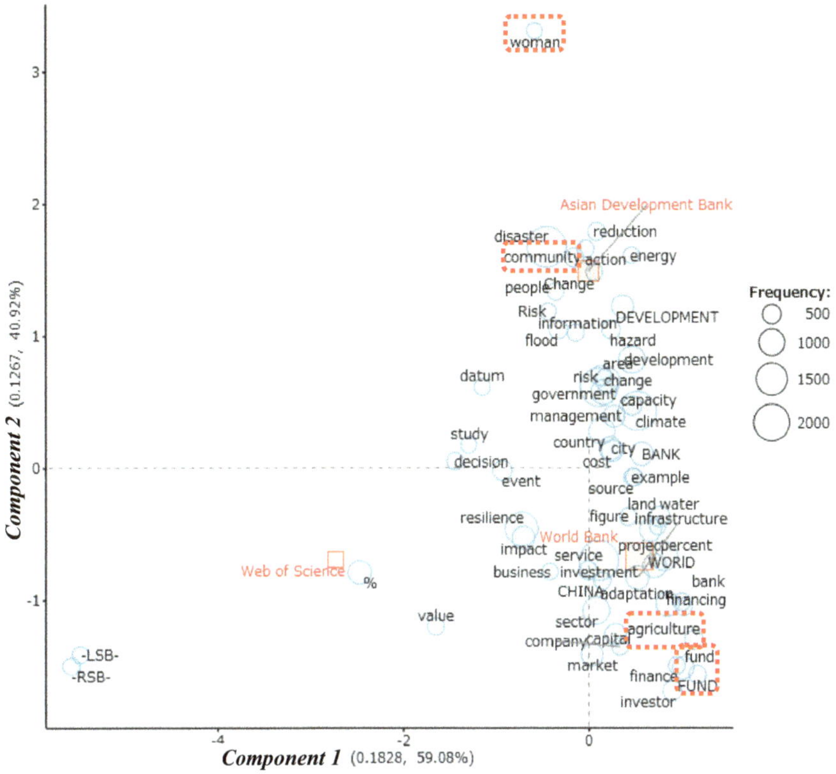

Fig. 3.3 The correspondence analysis. *Source* Sasaki (2021)

3.4 Conclusions

In this study, we reviewed the literature on DRR investments published since 2015, when the SFDRR was adopted. In the literature extracted from the databases of the World Bank and the ADB, some of the new concepts in DRR investments, such as the triple dividend of resilience and women-focused investment in resilience, have been presented. It is also pointed out that there exist some bottlenecks that hinder the promotion of investments in DRR, and the Blueprint for Action by the World Bank could serve as an example of a possible solution. Meanwhile, much of the literature extracted from the Web of Science are based on quantitative empirical analyses and provide a certain level of evidence for the concepts discussed in the World Bank and the ADB literature. In this context, the characteristic words of the literature extracted from the MDB databases through text mining can be deemed useful to a certain extent as a reference source when contemplating the possible direction of future academic research. The study reminds us that DRR investment is an extremely diverse topic in terms of disciplines and stakeholders. For academic research to contribute to society, it may be worthwhile to adopt a more transdisciplinary approach than ever before.

Acknowledgements This study was supported by the collaborative research project of the International Research Institute of Disaster Science, Tohoku University. This study was also partially supported by the Japan Society for the Promotion of Science KAKENHI, Grant Number 21H03680.

References

Asian Development Bank (2016) Incentives for reducing disaster risk in urban areas: experiences from Da Nang (Viet Nam), Kathmandu Valley (Nepal), and Naga City (Philippines)

Asian Development Bank (2020) Enhancing women-focused investments in climate and disaster resilience. https://doi.org/10.22617/TCS200140-2

Banerjee O, Boyle K, Rogers C, Cumberbatch J, Kanninen B, Lemay M, Schling M (2018) Estimating benefits of investing in resilience of coastal infrastructure in small island developing states: an application to Barbados. Mar Policy 90:78–87. https://doi.org/10.1016/j.marpol.2018.01.004

EU (2020) Regulation (Eu) 2020/852 Of The European Parliament And Of The Council of 18 June 2020 on the establishment of a framework to facilitate sustainable investment, and amending Regulation (EU) 2019/2088. http://data.europa.eu/eli/reg/2020/852/oj. Accessed 4 July 2021

Fung J, Helgeson J, Webb D, O'Fallon C, Cutler H (2020) Does resilience yield dividends? Co-benefits of investing in increased resilience in Cedar Rapids. Econ Syst Res 1–27. https://doi.org/10.1080/09535314.2020.1798359

Higuchi K (2016) A two-step approach to quantitative content analysis: KH Coder tutorial using Anne of Green Gables (Part I). Ritsumeikan Soc Sci Rev 52(3):77–91

Higuchi K (2017) A two-step approach to quantitative content analysis: KH Coder tutorial using Anne of Green Gables (Part II). Ritsumeikan Soc Sci Rev 53(1):137–147

ISO (2021a) ISO 14097:2021a. https://www.iso.org/standard/72433.html. Accessed 5 July 2021

ISO (2021b) ISO/FDIS 14030-1. https://www.iso.org/standard/43254.html. Accessed 5 July 2021b

ISO (2021c) ISO/FDIS 14030-2. https://www.iso.org/standard/75558.html. Accessed 5 July 2021c

ISO (2021d) ISO/DIS 14030-3.2. https://www.iso.org/standard/75559.html. Accessed 5 July 2021

ISO (2021e) ISO/FDIS 14030-4. https://www.iso.org/standard/76598.html. Accessed 5 July 2021

ISO (2021f) ISO/CD 14100. https://www.iso.org/standard/75619.html. Accessed 5 July 2021f

ISO (2021g) ISO/CD 32210. https://www.iso.org/standard/77776.html. Accessed 5 July 2021g.

Mechler R, Hochrainer-Stigler S (2019) Generating multiple resilience dividends from managing unnatural disasters in Asia: Opportunities for measurement and policy. Asian Development Bank Economics Working Paper Series, (601). https://doi.org/10.22617/WPS190573-2

Osti R (2017) Embedding community-based flood risk management in investment: a part-to-whole approach. Asian Development Bank. https://doi.org/10.22617/WPS179186-2

Overseas Development Institute, & World Bank (2015) Unlocking the 'Triple Dividend' of resilience: why investing in disaster risk management pays off. http://hdl.handle.net/10986/21612

Sasaki D (2021) Literature Review on investments in Disaster Risk Reduction. The Japan Society For International Development (JASID), June 2021 [Digests 22nd Spring Conference, 2021]

Tall A, Lynagh S, Vecchi C, Bardouille P, Pino F, Shabahat E, Stenek V, Stewart F, Power S, Paladines C, Neves P, Kerr L (2021) Enabling private investment in climate adaptation & resilience: Current status, barriers to investment and blueprint for action. World Bank

UN (2015) Resolution adopted by the General Assembly on 25 September 2015. https://www.un.org/en/development/desa/population/migration/generalassembly/docs/globalcompact/A_RES_70_1_E.pdf. Accessed 28 June 2021

UNDRR (2021) Commitment: Resilient Smart Community Infrastructure: Disaster Risk Reduction – Common Framework for Global Harmonization (ISO/TC268/SC1). https://sendaicommitments.undrr.org/commitments/20201225_001. Accessed 5 July 2021

UNFCCC (2016) Report of the Conference of the Parties on its twenty-first session, held in Paris from 30 November to 13 December 2015. https://unfccc.int/resource/docs/2015/cop21/eng/10.pdf. Accessed 3 July 2021
UNDRR (2015) Sendai framework for disaster risk reduction 2015–2030
World Bank (2016) Investing in urban resilience: protecting and promoting development in a changing world. World Bank. http://hdl.handle.net/10986/25219. Accessed 24 April 2021
Young K, Greenbaum R, Dormady N (2017) Sex, gender, and disasters: Experimental evidence on the decision to invest in resilience. Int J Disaster Risk Reduct 24:439–450. https://doi.org/10.1016/j.ijdrr.2017.06.003

Part II
Case Studies

Chapter 4
Exploring people's Perception of Disaster Risk Reduction Investment for Flood Management: The Case of Jakarta Floods in Indonesia

Dicky C. Pelupessy, Yasuhito Jibiki, and Daisuke Sasaki

Abstract This chapter examines how people perceive DRR investment for flood management in Jakarta, the capital of Indonesia and an Indonesian megacity. Rather than treating an individual's risk perception as a stand-alone concern, we assume that a person cannot be separated from their community and that their sense of community makes them aware of the risks involved while living in a disaster-prone area. We collected data through an online survey from March 2021 with 370 respondents residing in the Greater Jakarta area. The respondents were classified into four, gender-balanced age groups (20 s, 30 s, 40 s, and 50-and-over). Then, we first examine the correlation between the sense of community and the sense of resilience. Second, we verify whether the sense of resilience influences people's perceived need for flood investments. Multiple regression analysis identified that the sense of resilience affects investment need. Furthermore, two factors comprising the sense of resilience, place attachment and collective efficacy, influence investment need. The sense of community was weakly correlated with the sense of resilience. This implies that the sense of resilience may play a mediating role in the relationship between the sense of community and investment need.

Keywords Disaster Risk Reduction (DRR) investment for floods · Sense of community · Sense of resilience · Jakarta · Indonesia

D. C. Pelupessy (✉)
Faculty of Psychology, Universitas Indonesia, Depok, West Java, Indonesia
e-mail: dickypsy@ui.ac.id

Y. Jibiki · D. Sasaki
International Research Institute of Disaster Science (IRIDeS), Tohoku University, Sendai, Miyagi, Japan
e-mail: jibiki.yasuhito@gmail.com

D. Sasaki
e-mail: dsasaki@irides.tohoku.ac.jp

M. Ishiwatari and D. Sasaki (eds.), *Financing Investment in Disaster Risk Reduction and Climate Change Adaptation*, Disaster Risk Reduction,
https://doi.org/10.1007/978-981-19-2924-3_4

4.1 Introduction

4.1.1 Disaster Risk Reduction (DRR) Investment for Floods in Jakarta

In Indonesia, despite being an epicenter for floods, Jakarta is well known for having developed countermeasures for a long time. Some have noted that the history of flood protection in Indonesia dates back to the time of Tarumanegara Kingdom in the fifth century CE (Kompas 2021). During the Dutch colonial era, many efforts were made to tackle the impacts of flooding (Dwirahmadi et al. 2019).

Contemporarily, the Sendai Framework for Disaster Risk Reduction 2015–2030 (SFDRR) emphasizes on investment in DRR, including flood protection for increased resilience (Ishiwatari and Sasaki 2020; Ishiwatari and Sasaki 2021a, b). Even prior to the SFDRR's adoption, the Indonesian government had increased its flood protection budget by 30% since 2006 (Ishiwatari 2019; Ishiwatari and Sasaki 2021a, b). In the context of Jakarta, a World Bank report states that the completion of the East Flood Canal development contributed to a reduction in flooding in East and North Jakarta by approximately 30% (World Bank 2011). However, Hapsari and Zenurianto (2016) noted that the country at large still lacks sufficient flood prevention and preparedness measures. Jakarta is facing threats from complicated factors, including deforestation in the upstream areas, poor drainage systems, narrowing rivers and land subsidence associated with large-scale urbanization, and torrential rains due to climate change (Rahayu and Nasu 2010; Simanjuntak et al. 2012; Sagala et al. 2013; Budiyono et al. 2016; Mishra et al. 2018; Adibroto et al. 2019, Neise and Diez 2019; CNA 2021). Furthermore, over-extraction of groundwater has been recognized as main cause of land subsidence in Jakarta.

This study examines how people perceive DRR investment for flood management in Jakarta, the capital of Indonesia and an Indonesian megacity. Earlier works by Mechler (2005) and Hochrainer-Stigler et al. (2011) have focused on developing countries and mainly investigated the impact of disasters on investment, considering their damage size and characteristics, using a cost-benefit analysis. However, compared to this policymaking perspective, there are only a few empirical studies on people's perspectives as potential beneficiaries of DRR investment. In the Indonesian context, Putri et al. (2019) argued that current flood management practices tend to focus more on structural measures, while paying very little attention to the social processes involved in building a resilient society.

4.1.2 People's Perception of Floods in Jakarta

Numerous studies have investigated people's responses to floods in Jakarta. Among them, many researchers have paid considerable attention to those who live closer to rivers, including squatters, who are especially at risk. Texier (2008) and IRIDeS

(2013) noted that people continue to live their lives in flood-prone areas while being fully aware of the flood risk. Sunarharum et al. (2014) described that in Jakarta, such communities perceive flooding as a normal reality and a part of their daily lives.

Research suggests that people are less focused on the types of infrastructure developed against floods. For example, Marfai et al. (2015) showed that people believe that the lack of coordination and unsynchronized work is a bigger problem that should be prioritized over infrastructure development. Dwirahmadi et al. (2019) also revealed that people do not view the availability of flood walls and protective measures as a relevant characteristic of a flood-resilient community.

Some studies have highlighted community aspects, such as a sense of community and resilience. For example, Sunarharum et al. (2014) argued that people in Jakarta have already built their resilience by implementing adaptation strategies that have made them resistant to damage from flooding. Bott et al. (2019) empirically verified that the bonding ties of social capital are crucial to collective hazard response actions, and noted that an attachment to the place is an important factor for understanding hazard responses. Meanwhile, Texier (2008) and Van Voorst (2016) noted that state interventions did not necessarily succeed and that the local people did not trust these efforts. IRIDeS (2013) explained that many residing along Jakarta's rivers considered that garbage disposal caused flooding and blamed the government for failing to manage garbage clearance. These findings indirectly indicate a lower reliance on government efforts and emphasize a higher reliance on community resilience.

Our literature review revealed that while research has examined flood risk perception, investment need for flood, sense of community, and sense of resilience, these efforts seem fragmented. Here, we aim to contribute to the existing literature by assessing the relationship between these factors.

4.1.3 Hypothetical Framework

Rather than treating an individual's risk perception as a stand-alone concern, our study assumes that a person cannot be separated from their community and that their sense of community makes them aware of the risks involved while living in a disaster-prone area. Our hypothetical framework contains two steps: First, we examine the correlation between the sense of community and the sense of resilience (Fig. 4.1). Then, we verify whether the sense of resilience influences people's perceived need for flood investments. A detailed explanation of these concepts is provided in Sect. 4.2.2.

In the second step, other variables, such as the experience of floods, flood risk perception, and demographic variables, are controlled and compared using multiple regression analysis.

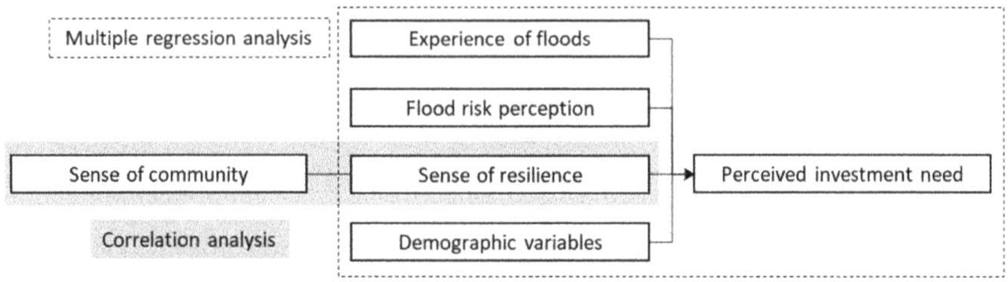

Fig. 4.1 Hypothetical framework for our analysis

4.2 Method

4.2.1 Participants

The study participants (N = 370) were adults aged 18–59 years (M = 37.7, SD = 10.5 years) residing in the Greater Jakarta area. Participants were selected using convenience sampling to ensure that we included 50% males and females to obtain a gender-balanced perspective on DRR investment. All participants provided informed consent before completing the online questionnaire.

4.2.2 Measures

To analyze people's perception of DRR investment, we utilize psychological indicators: the sense of community index (SCI) and the conjoint community resiliency assessment measure (CCRAM). SCI is a measure of sense of community (Perkins et al., 1990) based on McMillan and Chavis' (1986) theory of the sense of community. SCI does not specifically deal with disaster situations, while CCRAM clearly focus on disaster situations.

A sense of community constitutes the community members' feelings of belonging in terms of the belief that they matter to one another and to their community, and the shared faith that members' needs will be met through their commitment to live together (McMillan and Chavis 1986). A sense of community consists of five elements: membership, influence, integration, need fulfillment, and a shared emotional connection. SCI, which comprises 12 items with binary scoring (true or false), measures these five elements and enables us to comprehend how people perceive togetherness in their communities. To date, SCI is one of the most referenced measures of the sense of community concept, despite diverse views on the conceptualization of a sense of community. Here, we simply use the total sum of SCI.

CCRAM is a measure of community resilience. Leykin et al. (2013) developed CCRAM with the aim of establishing an integrated multidimensional instrument to assess community resilience. Later, Leykin et al. (2016) tested whether their measurement was robust as a psychological indicator. CCRAM shows people's evaluation of their communities' capacity to respond to emergencies in diverse contexts. It assesses the strength of five important factors of community function (leadership, collective efficacy, preparedness, place attachment, and social trust) that can be used to profile and predict community resilience (Cohen et al. 2013; Leykin et al. 2013, 2016).

CCRAM assesses the leadership factor in community resilience through six items representing general faith in decision makers, specific faith in local leaders, perception of fairness in the way local authority provides services, and functioning of the community. Collective efficacy is evaluated using five items representing collective efficacy, support, involvement in the community, and mutual assistance (Leykin et al. 2013). These five items echo the arguments of Norris et al. (2008) and Sampson et al. (1997) that collective efficacy comprises a composite of mutual trust and shared willingness to work for the common good of a neighborhood. The preparedness factor is evaluated using four items representing family and community acquaintance with emergency situations, and a view of the town's preparedness for emergency situations (Leykin et al. 2013). The place attachment factor is evaluated using four items representing emotional attachment to the community, sense of belonging, pride in the community, and ideological identification with the community (Leykin et al. 2013). This way of thinking about place attachment aligns well with Manzo and Perkins's (2006) assertion. Using a cross-disciplinary literature review, these authors noted that an individual's feelings toward a place are connected with community-level perceptions. Lastly, the social trust factor is evaluated using two items representing trust and quality of relationships between a community's members (Leykin et al. 2013).

Here, we used a 21-item self-reported CCRAM questionnaire. Each item was measured on a five-point Likert scale, with the answers providing the scores. The CCRAM total sum was calculated as the sum of all scores for each item. The five factors were generated following Leykin et al.'s (2013) categorization. For example, "leadership" consists of six items and its value was calculated as the sum of the scores of these six items. The 21-items of the CCRAM are listed in Table 4.5.

4.2.3 On-Line Survey Implementation

We conducted an online survey in March 2021 by recruiting 370 respondents residing in the Greater Jakarta area. The respondents were grouped into four age groups of people in their 20 s, 30 s, 40 s, and 50 s, and we allocated gender-balanced samples in each age group. For example, the over-50 age group consisted of 35 male and 35 female respondents.

A summary of demographic statistics is presented in Table 4.1. Regarding the respondents' primary occupation, for supplementary, 21.6% were retail sector employees, which was the largest proportion; 10.8% were homemakers and manufacturing sector employees, respectively. Others worked in construction, transportation, education, government, and other sectors.

Table 4.1 Demographic statistics summary

Variable	Percent	Variable	Percent
Gender		**Marital status**	
Male	50.0	Married	64.6
Female	50.0	Unmarried	31.1
Age (years)		Ever married	4.3
20–29	27.0	**Income level (annual)**	
30–39	27.0	< IDR 1,500,000	5.4
40–49	27.0	IDR 1,500,000– 2,499,999	5.7
> 50	18.9	IDR 2,500,000– 3,499,999	7.6
Education level		IDR 3,500,000–4,999,999	11.4
Elementary	1.4	IDR 5,000,000– 7,499,999	20.0
Junior High	3.0	IDR 7,500,000– 9,999,999	16.2
Senior High	27.3	> IDR 10,000,000 or more	33.8
Diploma 1, Diploma 2, Diploma 3, or Diploma 4	14.1		
University (Bachelor's degree)	47.6		
Post-graduate (Master's and PhD)	6.7		

Table 4.2. contains a summary of the basic statistics of SCI and CCRAM.

Table 4.2 Basic statistics of SCI and CCRAM

	Min	Max	Mean	SD
SCI	1	12	8.278	1.781
CCRAM total sum	23	105	74.797	12.511
CCRAM_1. Leadership	6	30	21.505	4.175
CCRAM_2. Collective efficacy	5	25	18.176	3.093
CCRAM_3. Preparedness	4	20	13.962	2.717
CCRAM_4. Place attachment	4	20	13.614	2.636
CCRAM_5. Social trust	2	10	7.541	1.385

SCI = Sense of community index; CCRAM = Conjoint community resiliency assessment measure; SD = Standard deviation

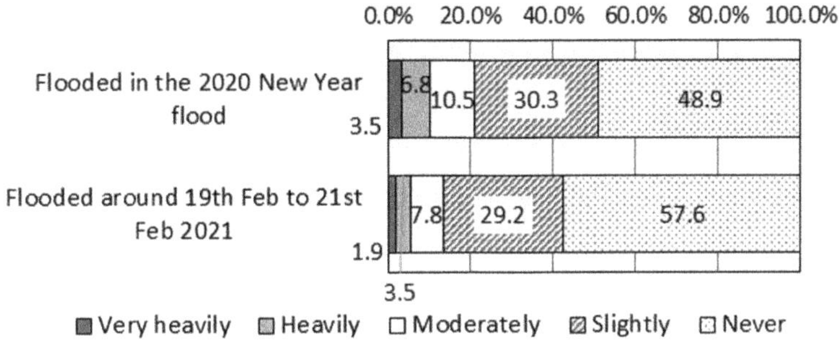

Fig. 4.2 Flood experience in recent flood events (n = 370)

4.3 Results

4.3.1 Experience of Flooding

First, we examine the respondents' flood experiences. We considered two significant flooding events that occurred recently before our data collection period: the 2020 New Year floods and the February 2021 flood. We asked the respondents whether they experienced flooding during either or both of these events. Almost half of the respondents reported experiencing a certain degree of flooding (Fig. 4.2). The magnitude of flooding appears to be relatively higher in the 2020 New Year flood than in the 2021 flood.

Notably, approximately 34% of respondents had not experienced flooding between 2016 and 2020. In contrast, less than 15% of respondents experienced floods more than five times in the past five years (see Fig. 4.3).

4.3.2 Flood Risk Perception and Investment Need

The respondents were asked whether their neighborhood was threatened by flood inundation in their daily life, rather than only during the wet season. The results showed that 10% of respondents perceived that their neighborhoods were threatened (upper bar in Fig. 4.4). While almost 70% of respondents perceived that their neighborhood was not threatened, 16.2% neither agreed nor disagreed that their neighborhood was threatened by flood inundation. We conducted a correlation analysis to clarify the relationship between flood risk perception and flood experience. The results showed a positive and statistically significant relationship between these two variables (Table 4.3). This implies that as the flood experience becomes more severe or frequent, the perceived need for investment is higher.

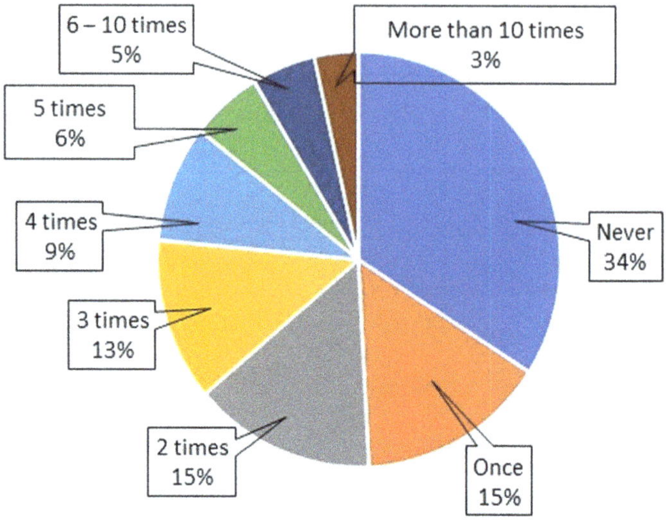

Fig. 4.3 Frequency of flood experiences (n = 370)

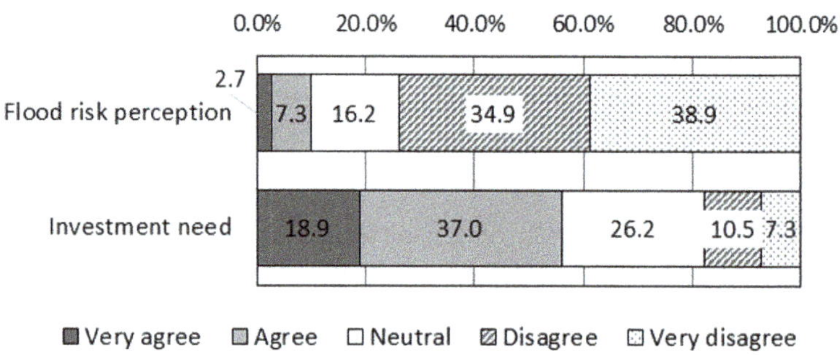

Fig. 4.4 Flood risk perception and investment need (n = 370)

Table 4.3 Correlation analysis between flood experience, risk perception, and investment need

	1	2	3	4	5
1. Flood experience: 2020 New Year					
2. Flood experience: 2021 February	0.718^{**}				
3. Frequency of Flood Experience: 2016–2020	0.644^{**}	0.609^{**}			
4. Flood risk perception in daily life	0.402^{**}	0.450^{**}	0.481^{**}		
5. Investment need	0.349^{**}	0.337^{**}	0.380^{**}	0.374^{**}	

Pearson product-moment correlation coefficient: **: $p < 0.01$.

Even though the respondents showed a low-risk perception, they exhibited a high perception of investment needs. Less than 20% respondents thought that flood investment should not be prioritized (lower bar in Fig. 4.4). The correlation between investment need, flood risk perception, and flood experience was positive and statistically significant (Table 4.3).

The results indicate:

- The more severe the flood experience, the higher the investment need.
- The higher the flood experience, the higher the investment need.
- The higher the day-to-day flood risk perception, the higher the investment need.

As our survey was carried out during the COVID-19 pandemic, we expected a relatively lower investment need because the respondents might consider investment in public health to be more important. However, our results differed from our assumption. The period in which we implemented the online survey was just after the Indonesian people had experienced the first wave of COVID-19 (around February 2021). At that time, the main agenda in public was to prepare for the anticipated second wave of COVID-19. Thus, we assumed that our respondents may have been influenced by such a situation and put their attention on public health topics rather than natural disasters.

4.3.3 Cross Tabulation Between Investment Need and the Demographic Variables

We tested whether the five demographic variables considered in this study had any relationship with investment need. Chi-square tests showed that there was a non-significant relationship between them (Fig. 4.5).

For simplification, we divided the respondents into two groups using an income level of five million IDR as the criteria. This is because the minimum wage in Jakarta in 2021 is approximately 4.4 million IDR per year; thus, 5 million IDR is a reasonable value. Regarding education, we integrated those who graduated from senior high schools, junior high schools, and elementary schools into one category ("Under senior high" in Fig. 4.5). Those who had master's and doctoral titles were combined into one group ("Post-graduate" in Fig. 4.5), with diploma holders and bachelor's degree holders forming the remaining two separate groups.

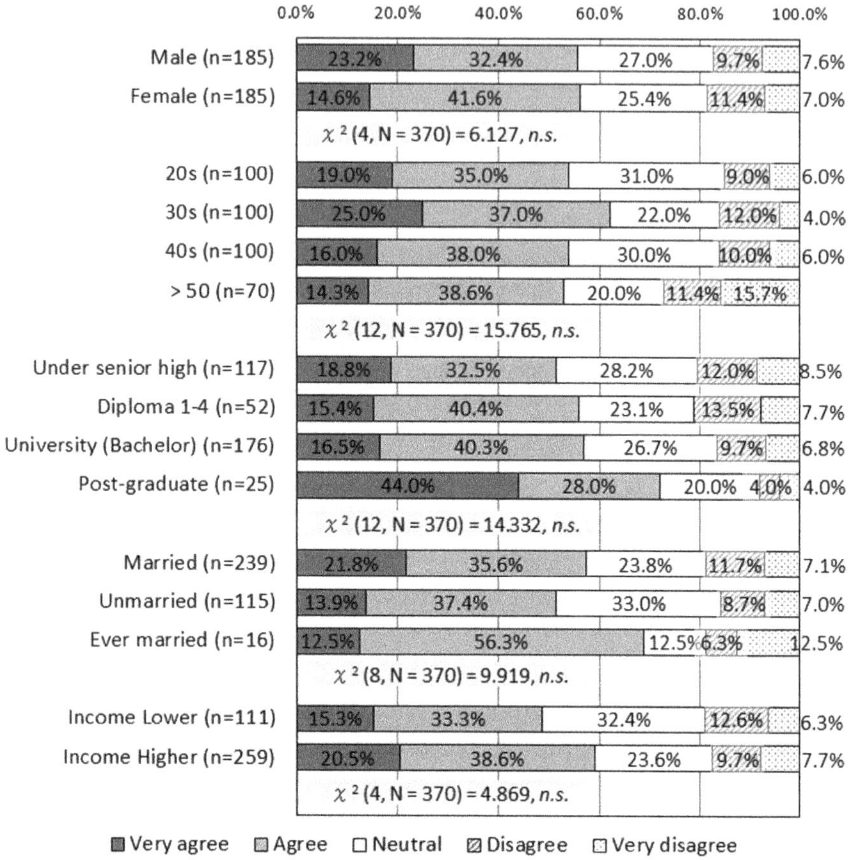

Fig. 4.5 Chi-square tests between investment need and demographic variables; n.s. denotes not significant

4.3.4 *Correlation Analysis Between Investment Need, SCI, and CCRAM*

We did not find a statistically correlated relationship between investment need and SCI (Table 4.4). This result supports our hypothesis that SCI does not directly influence DRR investment need.

Contrary to the SCI, we found a statistically significant, but weak relationship between investment need and CCRAM. The SCI had a significant relationship with the CCRAM total sum and the five factors of CCRAM. Furthermore, all CCRAM items had a significant relationship with SCI (Table 4.5). Such relationships could potentially imply that SCI was a basis or a latent factor behind the CCRAM. However, we found a weak correlation between the SCI and CCRAM ($r = 0.300$, $p < 0.01$, Table 4.4).

Table 4.4 Correlation analysis between investment need, SCI, and CCRAM

	1	2	3	4	5	6	7	8
1. Investment need								
2. SCI	−0.002							
3. CCRAM Total Sum	0.180**	0.300**						
4. CCRAM_1. Leadership	0.159**	0.242**	0.934**					
5. CCRAM_2. Collective Efficacy	0.176**	0.251**	0.929**	0.814**				
6. CCRAM_3. Preparedness	0.167**	0.272**	0.900**	0.823**	0.796**			
7. CCRAM_4. Place Attachment	0.150**	0.328**	0.816**	0.655**	0.719**	0.644**		
8. CCRAM_5. Social Trust	0.141**	0.257**	0.827**	0.740**	0.774**	0.678**	0.620**	

SCI = Sense of community index; CCRAM = Conjoint community resiliency assessment measure; Pearson product-moment correlation coefficient: **: $p < 0.01$.

4.3.5 *Multiple regression Analysis*

To identify variables that have a significant impact on investment needs, we tested five multiple regression analysis models. This method helped to control each independent variable and compare their degree of strength to explain the dependent variable. Table 4.6 provides a summary of the results.

All five models utilized nine independent variables: age, gender, marital status, education, household income, flood experience (two types: 2020 New Year and 2021 February), frequency of flood experience between 2016 and 2020, and flood risk perception. The CCRAM variables were also tested in three ways: total sum, five-dimension, and per-item-per-item.

In Models 1 and 2, we used the CCRAM total sum as an independent variable, but they differed in their statistical analytical methods: Model 1 used the forced imputation method, whereas Model 2 adopted the stepwise method.

In Models 3 and 4, the five CCRAM dimensions (leadership, collective efficacy, preparedness, place attachment, and social trust) were tested as independent variables. They differed in their statistical analytical method as follows: Model 3 used the forced imputation method, whereas Model 4 adopted the stepwise method.

In Model 5, we conducted a "per-item-per-item analysis" using the 21 items of the CCRAM. Because the 21 items were too many, we avoided running the forced imputation method for all these items, but only tested the stepwise method in Model 5. In the models that adopted the stepwise method, we showed only one result that had the largest adjusted R-square value in the results.

The results of Models 1 and 2 show that the total sum of CCRAM had a significant and positive impact on the dependent variable. This implies that as the total sum

Table 4.5 Correlation analysis between SCI and 21 items of CCRAM

	SCI
1. The municipal authority (regional council) of my town functions well [1: Leadership]	0.169**
2. There is mutual assistance and concern for others in my town [2: Collective Efficacy]	0.186**
3. My town is organized for emergency situations [3: Preparedness]	0.243**
4. I am proud to tell others where I live [4: Place Attachment]	0.274**
5. The relations between the various groups in my town are good [5: Social Trust]	0.262**
6. I have faith in the decision makers in the municipal authority (regional council) [1: Leadership]	0.221**
7. I can depend on people in my town to come to my assistance in a crisis [2: Collective Efficacy]	0.186**
8. The residents of my town are acquainted with their role is in an emergency situation [3: Preparedness]	0.232**
9. I feel a sense of belonging to my town [4: Place Attachment]	0.219**
10. There is trust among the residents of my town [5: Social Trust]	0.200**
11. In my town, appropriate attention is given to the needs of children [1: Leadership]	0.221**
12. There are people in my town who can assist in coping with an emergency [2: Collective Efficacy]	0.143**
13. In my town, there are sufficient public protection facilities (such as shelters) [3: Preparedness]	0.171**
14. I remain in this town for ideological reasons [4: Place Attachment]	0.155**
15. I have faith in the ability of the elected/nominated head of my town to lead the transit from routine to emergency management of the town [1: Leadership]	0.184**
16. I believe in the ability of my community to overcome an emergency situation [2: Collective Efficacy]	0.222**
17. My family and I are acquainted with the emergency system of my town (to be activated in times of emergency) [3: Preparedness]	0.241**
18. I would be sorry to leave the town where I live [4: Place Attachment]	0.341**
19. The municipal authority (regional council) provides its services in fairness [1: Leadership]	0.170**
20. The residents of my town are greatly involved in what is happening in the community [2: Collective Efficacy]	0.237**
21. The residents of my town will continue to receive municipal services during an emergency situation [1: Leadership]	0.249**

SCI = Sense of community index; CCRAM = Conjoint community resiliency assessment measure; Pearson product-moment correlation coefficient: **: $p < 0.01$.

Table 4.6. Results of multiple regression analysis

		Standardized Coefficient (p)	t Value	p Value	VIF		Standardized Coefficient (p)	t Value	p Value	VIF
Constant			2.593	0.010				2.575	0.010	
Age		−0.032	−0.644	0.520	1.160					
Gender		−0.019	−0.411	0.681	1.045					
Marital Status		0.005	0.102	0.919	1.216					
Education		0.065	1.226	0.221	1.311		0.090	1.967	0.050	1.001
Household Income		0.054	0.976	0.330	1.447					
Flood experience: 2020 New Year		0.131	1.S16	0.070	2.444		0.139	2.300	0.022	1.745
Flood experience: *2021* February		0.035	0.495	0.621	2.385					
Frequency of Flood Experience 2016–2020		0.162	2.463	0.014	2.047		0.178	2.914	0.005	1.901
Flood risk perception in daily life		0.216	3.982	0.000	1.382		0.221	4.182	0.000	1.330
CCRAM Total Sum		0.142	3.007	0.003	1.054		0.151	3.329	0.001	1.005
	R	0.4SS				R	0.485			
	adj R'	0.217				*adj R'*	0.224			
	N	370				N	370			
		Model 3: Forced inputation method					Model 4: Stepwisi e method			
		Standardized Coefficient (p)	t Value	p Value	VIF		Standardized Coefficient (p)	t Value	p Value	VIF
Constant			2.238	0.026				4.652	0.000	

(continued)

Table 4.6. (continued)

	Standardized Coefficient (p)	t Value	p Value	VIF	Standardized Coefficient (p)	t Value	p Value	VIF
Age	−0.033	−0.672	0.502	1.162				
Gender	−0.023	−0.487	0.626	1.060				
Marital Status	0.001	0.027	0.979	1.220				
Education	0.067	1.262	0.208	1.335				
Household Income	0.053	0.957	0.339	1.458				
Flood experience: 2020 New Year	0.136	1.859	0.064	2.501	0.135	2.223	0.027	1.745
Flood experience: 2021 February	0.035	0.487	0.627	2.401				
Frequency of Flood Experience 2016–2020	0.157	2.366	0.019	2.077	0.176	2.773	0.006	1.902
Flood risk perception in daily life	0.226	4.128	0.000	1.412	0.230	4.333	0.000	1.326
CCRAM 1. Leadership	−0.055	−0.571	0.568	4.397				
CCRAM 2. Collective Efficacy	0.063	0.629	0.530	4.717	0.153	3.319	0.001	1.003
CCRAM_3. Preparedness	0.015	0.165	0.869	3.807				
CCRAM 4. Place Attachment	0.077	1.123	0.262	2.227				
CCRAM 5. Social Trust	0.071	0.906	0.365	2.894				

(continued)

Table 4.6. (continued)

		Standardized Coefficient (p)	t Value	p Value	VIF		Standardized Coefficient (p)	t Value	p Value	VIF
	R	0.493				R	0.476			
	adj R'	0.213				$adj\ R^2$	0.219			
	N	370				N	370			
		Model 5: Stepwise method								
		Standardized Coefficient (p)	t Value	p Value	VIF					
Constant			2.321	0.021						
Education		0.096	2.108	0.036	1.014					
Flood experience: 2020 New Year		0.125	2.082	0.038	1.761					
Frequency of Flood Experience 2016–2020		0.175	2.796	0.005	1.910					
Flood risk perception in daily life		0.232	4.423	0.000	1.334					
CCRAM item no.14		0.109	2.223	0.027	1.176		CCRAM item no.l. 14:1 remain in this town for ideological reasors			
CCRAM item no.16		0.130	2.648	0.008	1.160		CCRAM item no.l6: I believe in the ability of my community to overcom e an emergency situation			
	R	0.501								
	$adj\ R^2$	0.239								
	N	370								

of CCRAM increases, their perception of investment need increases. The respondents evaluated the need for investment, even though they acknowledged a sense of resilience.

The results of Models 3 and 4 reveal that the collective efficacy factor has a significant and positive impact on the dependent variable. The other four factors were not statistically significant.

The results of Model 5 demonstrated that CCRAM item no. 14 (an item of place attachment) and CCRAM item no. 16 (an item of collective efficacy) both had significant and positive impacts on the dependent variable. Thus, as place attachment and collective efficacy increase, their perception of investment needs increases. That is, perceived investment need is higher when people have flooding experience and exhibit an attachment toward their place, and perceive higher collective efficacy.

Model 5 had the largest adjusted R-squared value among the five models. This means that Model 5 was the best among all models to explain perceived investment need. As previously mentioned, two items of the CCRAM were significant in Model 5. However, comparing the β values, the value for flood risk perception in daily life was larger. Moreover, in all models, the flood risk perception in daily life had the largest values of β.

4.4 Discussion

This study examined how people perceive DRR investment for flood management in Jakarta, Indonesia. Based on an online questionnaire survey, we used statistical analysis to show that sense of resilience affects people's perception of flood DRR investment needs.

Demographically, our respondents belonged to the middle-income category (Table 4.1). Furthermore, household income and education levels were relatively higher than average. In contrast, many articles have studied informal settlers in Jakarta, whose income and educational levels are relatively low. In addition, informal settlers tend to experience many floods and are highly exposed to flood risks. However, more recent flood occurrences show that the inundation is increasingly impacting middle-income neighborhoods.

We think that differences in the demographic features between our survey and other studies may have affected the results and interpretation of our analysis. For example, educational level has a significant effect on investment need. Meanwhile, the chi-square test did not show a significant relationship between educational level and investment need. Earlier studies have not clearly discussed the role of educational level. However, a multiple regression analysis in this study revealed that it was significant in comparison with the other variables.

We obtained new findings on flood experience, risk perception, and perceived investment need. Almost half of our respondents had few flood experiences (Fig. 4.3) and flood risk perception in their daily lives was relatively low (almost 10%; Fig. 4.4).

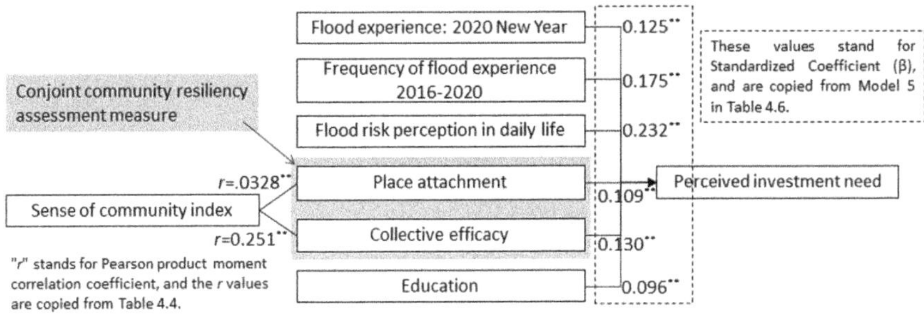

Fig. 4.6 Verifying our hypothetical framework

Contrary to these tendencies, the percentage of those who supported the need for investment reached nearly 60%.

The primary purpose of our study was to verify the hypothetical framework shown in Fig. 4.1. Figure 4.6 shows the verification results of this framework.

The sense of resilience, operationalized here as CCRAM scores, had a significant effect on perceived investment need. Compared with flood experience and flood risk perception in daily life, the explanatory power of the sense of resilience was not very strong. However, our analysis, particularly the multiple regression analysis, notably identified that a sense of resilience affects investment need.

The relationship between the sense of community, represented by SCI, and the sense of resilience, represented by CCRAM, showed a weak but significant correlation. This implies that the sense of resilience may play a mediating role between the sense of community and investment need. That is, the sense of resilience is connected to, but separated from, the sense of community. In this way, we see that the bond between community members, which orchestrates the sense of community, needs to be organized so that it can function for disaster risk reduction. This suggests the significance of the community in making an effort to organize action. That is, social action to respond to the challenges of disaster is organized in communities (Dynes 1998).

A deeper examination of the sense of resilience reveals that investment need is affected by place attachment and collective efficacy. Our results corroborate Bott et al.'s (2019) assertion about the importance of place attachment and that this factor needs to be further studied. We believe that our study extends the body of knowledge on people's perceptions of their community and DRR investment need. Further, we suggest that future studies should examine the role of place attachment and collective efficacy in DRR investment needs, specifically, and disaster risk reduction, in general.

Acknowledgements This study was supported by the Collaborative Research Project of the International Research Institute of Disaster Science, Tohoku University. This study was also partially supported by the Japan Society for the Promotion of Science KAKENHI, Grant Number 21H03680.

References

Adibroto TA, Wijayanti P, Adhi RP, Nugroho R (2019) Preliminary study on socio-economic aspect towards Jakarta climate resilient (case study: Cengkareng District, West Jakarta and Penjaringan District, North Jakarta). In: IOP conference series: earth and environmental science, vol 314, No. 1, p 012035). IOP Publishing. https://iopscience.iop.org/article/https://doi.org/10.1088/1755-1315/314/1/012035/pdf. Accessed 10 Sept 2021

Bott LM, Ankel L, Braun B (2019) Adaptive neighborhoods: the interrelation of urban form, social capital, and responses to coastal hazards in Jakarta. Geoforum 106:202–213

Budiyono Y, Aerts JC, Tollenaar D, Ward PJ (2016) River flood risk in Jakarta under scenarios of future change. Nat Hazard 16(3):757–774

CNA (Channel News Asia) (2021) The fight against Jakarta's devastating yearly floods◦https://www.channelnewsasia.com/news/asia/jakarta-flooding-relocation-giant-seawall-anies-baswedan-13864908. Accessed 10 Sept 2021

Cohen O, Leykin D, Lahad M, Goldberg A, Aharonson-Daniel L (2013) The conjoint community resiliency assessment measure as a baseline for profiling and predicting community resilience for emergencies. Technol Forecast Soc Chang 80(9):1732–1741

Dwirahmadi F, Rutherford S, Phung D, Chu C (2019) Understanding the operational concept of a flood-resilient urban community in Jakarta, Indonesia, from the perspectives of disaster risk reduction, climate change adaptation and development agencies. Int J Environ Res Public Health 16(20):3993

Dynes R (1998) Coming to terms with community disaster. In: Quarantelli EL (ed) What is a disaster? Perspectives on the questions. Routledge, New York, pp 109–126

Hapsari RI, Zenurianto M (2016) View of flood disaster management in Indonesia and the key solutions. Am J Eng Res 5(3):140–151

Hochrainer-Stigler S, Kunreuther H, Linnerooth-Bayer J, Mechler R, Michel-Kerjan E, Muir-Wood R, Ranger N, Vaziri P, Young M (2011) The costs and benefits of reducing risk from natural hazards to residential structures in developing countries. (Working Paper # 2011–01, The Wharton School, University of Pennsylvania)

IRIDeS (International Research Institute of Disaster Science) (2013) Second REPORT of IRIDeS Fact-finding mission to Jakarta, Indonesia. https://irides.tohoku.ac.jp/media/files/topics/irides_jakarta_2nd_report.pdf. Accessed 10 Sept 2021

Ishiwatari M (2019) Investing in disaster risk reduction: scale and effect of investment in flood protection in Asia. Contributing Paper to GAR 2019

Ishiwatari M, Sasaki D (2020) Bridging the gaps in infrastructure investment for flood protection in Asia (No. 202). (JICA Ogata Research Institute Working Paper No.202)

Ishiwatari M, Sasaki D (2021a) Investments in flood protection: Trends in flood damage and protection in growing Asian economies. (JICA Ogata Research Institute Working Paper No.221)

Ishiwatari M, Sasaki D (2021b) Investing in flood protection in Asia: an empirical study focusing on the relationship between investment and damage. Prog Disaster Sci 12:100197

Kompas (2021) Sejarah Banjir Jakarta dari Zaman Tarumanegara hingga Hindia Belanda (The History of the Jakarta Flood from the Tarumanagara Period to the Dutch East Indies). https://www.kompas.com/skola/read/2021/02/22/084300069/sejarah-banjir-jakarta-dari-zaman-tarumanegara-hingga-hindia-belanda?page=all. Accessed 10 Sept 2021

Leykin D, Lahad M, Cohen O, Goldberg A, Aharonson-Daniel L (2013) Conjoint community resiliency assessment measure-28/10 items (CCRAM28 and CCRAM10): a self-report tool for assessing community resilience. Am J Community Psychol 52(3–4):313–323

Leykin D, Lahad M, Cohen R, Goldberg A, Aharonson-Daniel L (2016) The dynamics of community resilience between routine and emergency situations. Int J Disaster Risk Reduct 15(1):125–131

Manzo LC, Perkins DD (2006) Finding common ground: the importance of place attachment to community participation and planning. J Plan Lit 20(4):335–350

Marfai MA, Sekaranom AB, Ward P (2015) Community responses and adaptation strategies toward flood hazard in Jakarta, Indonesia. Nat Hazards 75(2):1127–1144

McMillan DW, Chavis DM (1986) Sense of community: a definition and theory. J Community Psychol 14(1):6–23

Mishra BK, RafieiEmam A, Masago Y, Kumar P, Regmi RK, Fukushi K (2018) Assessment of future flood inundations under climate and land use change scenarios in the Ciliwung River Basin, Jakarta. J Flood Risk Manag 11:S1105–S1115

Mechler R (2005) Cost-benefit analysis of natural disaster risk management in developing countries. Working paper for sector project "Disaster Risk Management in Development Cooperation", GTZ

Neise T, Diez JR (2019) Adapt, move or surrender? Manufacturing firms' routines and dynamic capabilities on flood risk reduction in coastal cities of Indonesia. Int J Disaster Risk Reduct 33:332–342

Norris FH, Stevens SP, Pfefferbaum B, Wyche KF, Pfefferbaum RL (2008) Community resilience as a metaphor, theory, set of capacities, and strategy for disaster readiness. Am J Community Psychol 41(1):127–150

Perkins DD, Florin P, Rich RC, Wandersman A, Chavis DM (1990) Participation and the social and physical environment of residential blocks: Crime and community context. Am J Community Psychol 18(1):83–115

Putri IAP, Dalimunthe SA, Prasojo APS (2019) The right to live dangerously: public perceptions of extreme water events in urban areas. In: 2nd international conference on strategic and global studies (ICSGS 2018), vol 365. Atlantis Press, The Netherlands

Rahayu HP, Nasu S (2010) Good practices of enhancement early warning system for high populated cities: a case study for Jakarta flood. Soc Soc Manag Syst Internet J 6 (1). http://hdl.handle.net/10173/1827

Sagala S, Lassa J, Yasaditama H, Hudalah D (2013) The evolution of risk and vulnerability in Greater Jakarta: contesting government policy in dealing with a megacity's exposure to flooding, pp 1–18. (Working Paper; No. 2). Institute of resource governance and social change. http://irgsc.org/pubs/wp/IRGSCWP002jakartaflood.pdf. Accessed 19 Jan 2022

Sampson R, Raudenbush S, Earls F (1997) Neighborhoods and violent crime: a multilevel study of collective efficacy. Science 277:918–924

Simanjuntak I, Frantzeskaki N, Enserink B, Ravesteijn W (2012) Evaluating Jakarta's flood defense governance: the impact of political and institutional reforms. Water Policy 14(2012):561–580

Sunarharum TM, Sloan M, Susilawati C (2014) Community engagement for disaster resilience: flood risk management in Jakarta, Indonesia. Proc Second ANDROID Residlal Dr Sch Work Packag III:151–160

Texier P (2008) Floods in Jakarta: when the extreme reveals daily structural constraints and mismanagement. Disaster Prev Manag: Int J 17(3):358–372

Van Voorst R (2016) Formal and informal flood governance in Jakarta, Indonesia. Habitat Int 52:5–10

World Bank. (2011) Jakarta - Urban challenges in a changing climate (English). Washington, D.C.: World Bank Group. http://documents.worldbank.org/curated/en/132781468039870805/Jakarta-Urban-challenges-in-a-changing-climate. Accessed 10 Sept 2021

Chapter 5
Factors Influencing Climate Change Adaptation Investment by Local Government Units in the Philippines

Kaori Shiohara

Abstract Local governments are on the frontline of planning and implementing climate change adaptation (CCA). However, knowledge of the status and determining factors of local level public investment in CCA is limited. This study examined approaches to monitor local government investments in CCA. It attempted to (1) understand the state of public investment in CCA by local governments in the Philippines and (2) examine the determining factors of this investment using local climate change expenditure tagging data from the 2016 Annual Investment Programs of Philippine local government units (LGUs). The data were documented by and obtained from the Climate Change Commission, a policymaking body of the Philippine government. The Philippines has experienced climate-and water-related disasters and is expected to become more vulnerable to climate change. The country has been prompt in institutional and legal responses to climate change. The results show that CCA investment ratios are not related to the fiscal scale of LGUs. No clear link between CCA investments and flood risk was evident. Rather, the study suggests that central-local government relationships, the scope of CCA measures, and political connections affect CCA investment by LGUs.

Keywords Climate change adaptation · Local government · Philippines · Monitoring

5.1 Introduction

The Paris Agreement was adopted in 2015 as an international treaty on climate change. It set global targets for climate change adaptation (CCA). As established in the Paris Agreement goals, each country should work to develop adaptation plans and implement programs. Developing nations should also play a part in these adaptation efforts. International policies have strongly influenced developing countries in

---

K. Shiohara (✉)
The University of Tokyo, Tokyo, Japan
e-mail: kaori.shiohara1701@gmail.com

M. Ishiwatari and D. Sasaki (eds.), *Financing Investment in Disaster Risk Reduction and Climate Change Adaptation*, Disaster Risk Reduction,
https://doi.org/10.1007/978-981-19-2924-3_5

incorporating climate change initiatives into their development planning (Sova and Schipper 2019).

The Philippines is one of the countries with the fastest institutional responses to climate change (Raquedan 2010). Its official policies and institutional measures regarding climate change date back to the May 1991 Inter-Agency Committee on Climate Change (Raquedan 2010). This response arose from a sense of urgency, as the Philippines is highly vulnerable to the effects of climate change (Republic of the Philippines 1991). The Climate Change Act, enacted in October 2009, was the first of its kind in developing countries in terms of comprehensiveness (Oulu 2015).

The Climate Change Act aims to make climate change a mainstream component of policy formation, develop a strategic framework and programs to counter climate change, and establish the Climate Change Commission (CCC; Republic of the Philippines 1991) as a policymaking body of the Philippine government. It also clearly emphasizes adaptation measures (de Leon and Pittock 2017). Additionally, the Act positions local government units (LGUs), including provinces, cities, municipalities, and barangays (administrative units smaller than cities and municipalities) as frontline entities to formulate, plan, and implement climate policy in their respective jurisdictions.

In 2014, the Philippine government integrated climate change expenditure tagging (CCET) into its national budget. In January 2015, the Department of Budget and Management (DBM) and the CCC issued the DBM-CCC Joint Memorandum Circular 2015–01 "Revised Guidelines for CCET amending JMC No. 2013–01." The guidelines required all governmental entities to identify, plan, report, and track their climate budgets (CCC and DBM 2016). Local Budget Memorandum No. 70 was also issued in the same year, making climate budgets mandatory for LGUs (Andreas et al. 2018). Thus, in addition to being proactive regarding CCA, the Philippines is an important research subject because it incorporates climate change initiatives into its budgetary process.

This study examines approaches to monitoring local government investment in CCA through a case study of the Philippines. The study first attempted to understand the current status of CCA investment by LGUs in the Philippines. It then examined the determining factors of this investment. In light of the above, this study poses two research questions. First, how do LGUs in the Philippines invest in CCA in their annual investment programs. Second, what factors are involved in determining the amounts and ratios of CCA investment?

The next section reviews previous studies on adaptation finance and related literature on the factors determining CCA investment by local governments.

5.2 Literature Review

Prior research on adaptation finance has mainly focused on international funding; there is scant subnational level research. The existing literature on adaptation finance focuses primarily on international funding or the international distribution of climate change finance (Barrett 2014; Pannier et al. 2020), especially the Adaptation Fund under the United Nations Framework Convention on Climate Change (Barrett 2015). However, when considering the issue of CCA finance as a whole, the discussions concentrated on fund allocation at international and national levels. The lack of discussion on public investment, especially by local governments, which are the main planning and implementation factors for CCA, overlooks a key element in understanding adaptation finance. Therefore, research on CCA investments at the subnational level should be encouraged.

Research on CCA investment determinants, particularly those concerning local governments, is limited. Recognizing that no large-N research exists that addresses the determining factors of adaptation finance, Barrett (2014) discussed the subnational allocation of adaptation finance in Malawi. However, the main subject of this study is aid. Although Karim and Noy (2020) discussed the determinants of public expenditures on disaster risk reduction (DRR) and CCA at the subnational level, the only relevant study, as discussed below, is that by Miller and Vela (2014), according to Karim and Noy (2020).

Recent studies have found that factors influencing CCA investment by LGUs include risks associated with climate, flooding, and poverty rate. Karim and Noy (2020) examined central government funding allocations to local governments in DRR and CCA in Bangladesh and showed that funding allocations to subnational levels could be explained by flood hazard risk, poverty rate, and other factors indicating socioeconomic vulnerability. Musah-Surugu et al. (2018), who investigated the extent to which local governments in Ghana adhered to guidelines for budget compilation and climate change finance, showed that exposure to local level climate change significantly influenced the willingness to comply with legal authorities. Based on their results, they predicted that local government bodies highly vulnerable to climate change are more likely to allocate significant resources to adaptation-related initiatives than those that are not (Musah-Surugu et al. 2018).

The few studies that have discussed DRR at the subnational level have identified several determinants that have implications for CCA investment. Miller and Vela (2014) aimed to identify the factors affecting regional allocations of public spending on DRR in Peru. The authors asserted that DRR public spending can be largely explained by hazard risk based on long-term historical exposure. Further, Islam et al. (2021) showed that political-economic factors are key to DRR funding allocation at the sub-district level in Bangladesh.

These determinants have also been discussed in studies of environmental expenditure at the local level. According to Broietti et al. (2018), prior studies have identified factors such as population, gross domestic product, per capita income, total area, the state's fiscal health, and intergovernmental relations. The authors argued that these factors can be classified into specific categories, such as socioeconomic and structural factors, based on Bacot and Dawes (1997).

In particular, Bacot and Dawes (1997) are cited frequently. The authors contended that population is clearly "the overwhelming determinant" of state-level environmental expenditures. The basic thinking behind this contention is that states with large populations have the most fiscal and political power, and are therefore most able to enforce regulations mandated by the central government (Bacot and Dawes 1997). Thus, it can be surmised that the population and financial scale that result from having a large population can also define the environmental investment of local governments.

5.3 Methods

This study used CCA investment data from the 2016 annual investment programs of 130 LGUs in the Philippines collected by CCC (n.d.). Annual investment programs use the CCET system as a basis for tagging investment in CCA and climate change mitigation. Local CCET is a process linking programs, projects, and activities (PPAs) listed in LGUs' annual investment programs related to CCA or climate change mitigation with typologies of climate change activities. Typologies are categorized by sectors and measures that include "Policy Development and Governance," "Research, Development and Extension," "Knowledge Sharing and Capacity Building," and "Service Delivery". Codes are then assigned to them (DBM et al. n.d.). Through CCET, the investment amount planned for climate change measures can be visualized for each typology. Of the total climate change investment in 2016 by the 130 LGUs, 98.2% were for CCA and 1.8% were for climate change mitigation. Investment tagged as either CCA or climate change mitigation constituted 28.8% of the total investment of 130 LGUs in 2016 (CCC n.d.).

Correlation analysis was performed using Stata software to identify the factors explaining the differences in the amounts and ratios of CCA investment among different LGUs. Exploratory measurement of the degree to which each variable could or could not be related to the scale and ratio of CCA investments was considered necessary. Therefore, a correlation analysis was conducted. Based on the literature review, the following variables were used as potential explanatory variables for the analysis: the aforementioned data on the LGU's total annual investment program, which indicates each LGU's scale of total investment; the total population in each LGU (Philippine Statistics Authority [PSA] 2015c); LGU income (Department of Finance 2015); the poverty rate in LGUs as an indicator of social vulnerability (PSA 2015a, 2015b); and the average flood risk, which appears to be related to physical vulnerability (Mines and Geosciences Bureau [MGB] n.d.). The inclusion of physical vulnerability was also based on the reviewed literature, which highlighted that the risks of climate-related impacts could explain CCA investment by local governments.

The average flood risk is based on data concerning susceptibility to flood damage in the MGB's (n.d.) hazard susceptibility rating database available on the website of the Department of Environment and Natural Resources, Mines and Geosciences Bureau (DENR-MGB). The database ranks flood susceptibility at the barangay level on a 5-point susceptibility scale as follows: 0, "None"; 1, "Low Susceptibility to Flooding"; 2, "Moderate Susceptibility to Flooding"; 3, "High Susceptibility to Flooding"; and 4, "Very High Susceptibility to Flooding". The average flood susceptibility within each province, city, and the municipality was calculated and used as the index. This study also referred to the World Food Programme (2014) to examine the methodology for creating this study's index.

Additional data were used to investigate these results further. The financing sources for CCA investment by local governments were based on the funding source categorizations used by the CCC (n.d.). Local climate budgets financed by the national government were aggregated according to the categories of "National Government", "National Government Other External Funding Sources", and "National Government Agencies". The co-financing of the national government and LGUs was aggregated into the "LGU–National Government" and "LGU–National Government Other External" Categories. CCET data at the national government level were obtained from the CCC (2016), which provides a compilation and analysis of CCET data of national government agencies in the Philippines for the fiscal year 2016.

5.4 Results

The current status of CCA investment was examined. In terms of the relationships between relevant national entities and LGUs in mainstreaming local level climate policy, the CCC provides LGUs and others with funding specifically for CCA projects through the People's Survival Fund (PSF), a special fund supported by the General Appropriations Act. With the exception of DRR, PSF seems to be a limited route through which funds for additional CCA projects can be applied for and obtained from the central government by local governments.

The PSF was established in August 2012 under the provisions of Republic Act 10174. The decision was later made to allocate one billion PHPs (approximately 20 million USD) per year to the PSF. The purpose of the fund is to provide financing for adaptation activities by local governments and accredited local community organizations (Government of the Philippines n.d.). Since its inception, the fund has approved six projects as of 2019. Four of these designated project areas are in the Mindanao region, including the provinces of Surigao del Norte and Surigao del Sur (Thailand Development Research Institute [TDRI] 2019), which have a higher than average ratio of CCA investments.

Table 5.1 ranks the PPAs tagged as climate change investments of the 130 LGUs in their annual investment programs submitted to the CCC in the 2016 fiscal year. The top five account for approximately 80% of the total investment and focus mainly on hard infrastructure. This composition is probably one of the reasons CCA investment comprises 98.2% of total climate change investment. In order of frequency, the most common PPAs focus on upgrading existing highways, ports, and airport infrastructure to climate change-resilient design standards (1,175 projects), construction of new highways and port infrastructure to climate change-resilient design standards, and countermeasures and reconstruction projects aimed at protecting government infrastructure from the hazards of climate change (CCC n.d.). Table 5.1 is based on

Table 5.1 Top five programs, projects, and activities tagged as climate change investment by local government units by total expenditures in the 2016 fiscal year

Rank	Typology based on climate change expenditure Tagging	% of total climate change expenditure (%)
1	"Upgrade existing roads, ports, and aviation infrastructure to climate-resilient design standards"	32
2	"Construct new roads, ports, airports, and aviation infrastructure to climate-resilient design standards"	21
3	"Improve the resilience of infrastructure (bridges, water supply, community infrastructure, water storage, coastal defense, etc.) to account for climate change and climate variability-related extreme weather, and climate variability that could increase flood risks in infrastructure"	17
4	"Incorporate climate change and climate variability in design standards for flood control and drainage systems"	6
5	"Climate-proofing/retrofitting or relocating government infrastructure (i.e. schools and government hospitals, health centers, rural health units, teen centers) from climate hazards"	4

The table was adapted from the Climate Change Commission (n.d.)

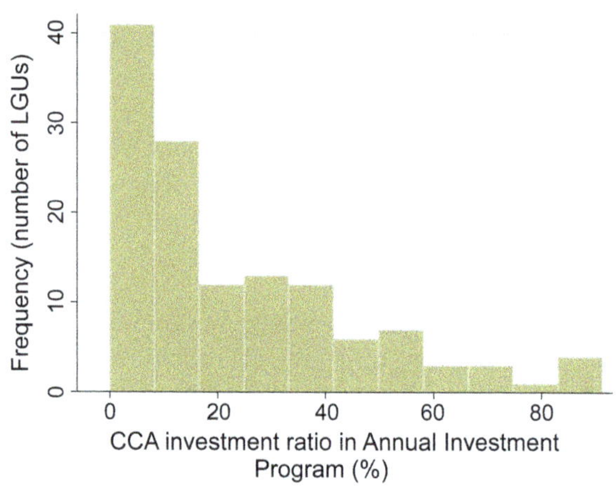

Fig. 5.1 Climate change adaptation (CCA) investment ratio in the Annual Investment Programs of 130 local government units (LGUs; provinces, cities, municipalities) in the 2016 fiscal year. The calculation of the ratio of CCA investment to total investment is based on the Annual Investment Programs of the 130 LGUs (CCC n.d.)

typology codes for the CCET system, which may differ from the actual names of PPAs listed in the annual investment programs.

Figure 5.1 shows the distribution of CCA investments across the 130 LGUs (mean ± standard deviation, 23.46 ± 22.17). Among the 130 LGUs studied, 22 were provinces, 12 were cities, and 96 were municipalities. The CCA investment ratio indicates the percentage of the total investment in LGUs' annual investment in CCA activities. The X-axis shows the CCA investment ratio of each LGU and the Y-axis shows the number of LGUs that fall within a specific investment ratio.

Table 5.2 displays the basic information of the nine provinces for which the ratio of CCA investment is higher than the average for the LGUs studied.

Next, the factors affecting the ratio and amount of CCA investments were investigated. Correlation analysis revealed that none of the indices of the scale of total investment in the annual investment program, total population, LGU income, or poverty rate correlated with the CCA investment ratio for each LGU (Table 5.3, Fig. 5.2). Regarding the amount of CCA investment by each LGU, the amount of CCA investment was not correlated with the total population, LGU income, and poverty rate (Table 5.4). The fiscal scale at the local level (the scale of the total investment in the annual investment program) showed no correlation with the CCA investment ratios. In contrast, fiscal scale correlated with the amount of CCA investment, with a coefficient variable of 0.6022 (Table 5.4, Fig. 5.3). Furthermore, this study could not attribute the differences between LGUs' CCA investments and flood

Table 5.2 Nine provinces for which the ratio of climate change adaptation (CCA) investment is higher than the average

Ranking	Region	Province	CCA investment ratio (%)	Amount of CCA investment (in PHP '000) (1 PHP = 0.02 USD)
1	Region V	Camarines Sur	90.68	10,006,348 (200,127 USD)
2	Region XIII	Surigao del Sur	75.04	4,091,290 (81,826 USD)
3	Region XIII	Surigao del Norte	73.34	9,174,048 (183,481 USD)
4	Region V	Albay	63.57	27,618,642 (552,373 USD)
5	Region III	Bulacan	38.44	3,202,495 (64,050 USD)
6	Region III	Nueva Ecija	36.56	3,638,780 (72,776 USD)
7	Region V	Sorsogon	33.08	2,911,662 (58,233 USD)
8	Region XII	North Cotabato	24.13	784,312 (15,686 USD)
9	Region VII	Bohol	24.07	639,791 (12,796 USD)

CCA investment ratio was calculated by the author (Climate Change Commission, n.d.)

Table 5.3 Correlation analysis of local government unit (LGU) climate change adaptation (CCA) investment ratios and relevant indices

	CCA investment ratio	Scale of total investment	Total population	LGU income	Poverty rate
CCA investment ratio	1				
Scale of total investment	0.0749	1			
Total population	0.021	0.402	1		
LGU income	−0.006	0.3825	0.8285	1	
Poverty rate	0.1175	−0.08	0.0027	0.0462	1

The sample size is 130

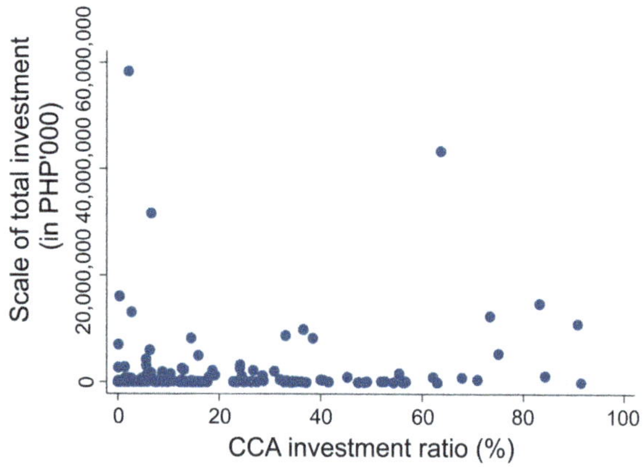

Fig. 5.2 Relationship between the scale of total investment in annual investment programs and climate change adaptation (CCA) investment ratios in local government units

Table 5.4 Correlation analysis of the total amounts of local government unit (LGU) climate change adaptation (CCA) investment and relevant indices

	Amount of total CCA investment	Scale of total investment	Total population	LGU income	Poverty rate
Amount of total CCA investment	1				
Scale of total investment	0.6022	1			
Total population	0.301	0.402	1		
LGU income	0.2813	0.3825	0.8285	1	
Poverty rate	0.0302	−0.08	0.0027	0.0462	1

The sample size is 130

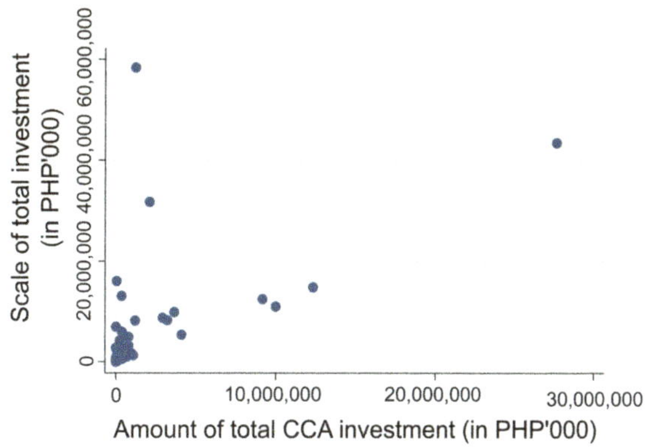

Fig. 5.3 Relationship between the scale of total investment in annual investment programs and the amount of total climate change adaptation (CCA) investment by local government units

risk, as no correlation was found between flood risk and CCA investment amounts and ratios (Table 5.5, Figs. 5.4 and 5.5).

Instead of the variables examined in the correlation analysis, three other factors may increase LGUs' CCA investment. The first is the differences in financing sources among individual LGUs for CCA programs that reflect the relationship between

Table 5.5 Correlation analysis of the amount and ratio of local government unit (LGU) climate change adaptation (CCA) investment and flood risk

	Amount of total CCA investment	CCA investment ratio	Flood risk
Amount of total CCA investment	1		
CCA investment ratio	0.4443	1	
Flood risk	−0.0533	0.1544	1

The sample size is 94

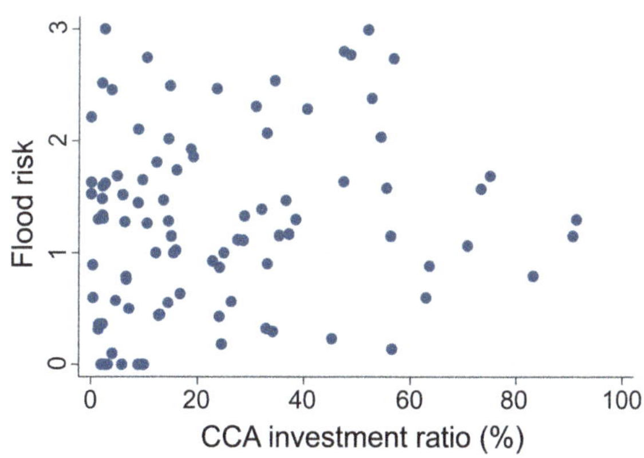

Fig. 5.4 Relationship between flood risk and climate change adaptation (CCA) investment ratios in local government units

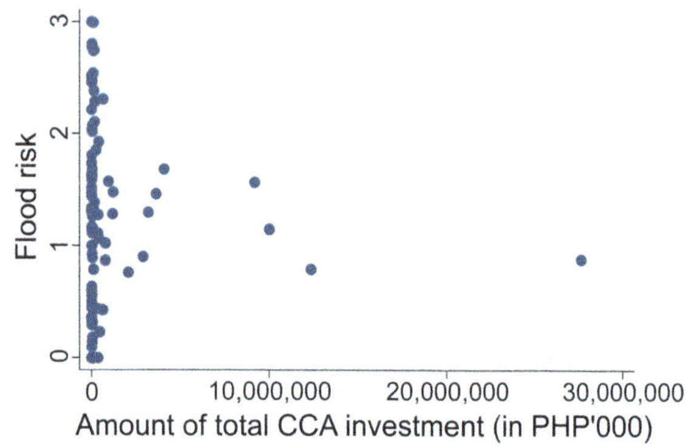

Fig. 5.5 Relationship between flood risk and the amount of total climate change adaptation (CCA) investment by local government units

national and local governments. The second is the scope of CCA measures. The third is political connections. Based on further analysis of the study results, this section explains the first and second factors. The third is explained in the next section.

Different sources of financing among LGUs for CCA programs may affect their CCA investments. This is because, among the provinces, those with an above-average CCA investment ratio tend to have a higher proportion of contribution from the national government or joint contributions from the national and local governments for their CCA program financing. Here, funds allocated directly to LGUs include contributions from the national government or co-financing by national and local governments according to the aforementioned funding source categorizations used by the CCC (n.d.). For example, according to the funding source breakdown for climate change investment, the funding for total climate investment (of which 98.2% was for CCA investment) that came from national government contributions and/or co-financing calculated by the author based on CCC [n.d.] data was as follows: Camarines Sur at 99% (30% national government, 69% co-financing), Surigao del Sur at 90% (79% national government, 11% co-financing), Surigao del Norte at 82% (69% national government, 13% co-financing), Albay at 50% (1% national government, 49% co-financing), Nueva Ecija at 89% (co-financing), and Sorsogon at 90% (50% national government, 40% co-financing). Other LGUs tend to use special funds and domestic resources to finance their main CCA measures. Excluding the provinces with above-average CCA investment ratios, the rest of the provinces averaged only 14.9% of their funding for climate investment from national government contributions and/or co-financing (calculated by the author based on CCC [n.d.] data). Therefore, a clear difference exists regarding the financing sources used by provinces with above-average CCA investment ratios versus other provinces.

An example of funds contributed by the central government is distributed through priority programs developed and implemented by national government agencies as part of climate policy. Among national government agencies, the main contributors to the central government's climate investment are the Department of Public Works and Highways (DPWH), Department of Agriculture (DA), and Department of Environment and Natural Resources (DENR). The DPWH primarily structures its climate budget through PPAs related to flood management and national road networks. The other departments have typically incorporated key programs into their climate budgets, with the DENR running the National Greening Program (comprising 61% of the department's climate budget in the fiscal year 2016) and the DA responsible for the Philippine Rural Development Program (constituting 51% of the department's climate budget in 2016 fiscal year) (CCC 2016).

The ability of each LGU to access climate-related funding from the DPWH likely considerably impacts both its CCA investment amount and ratio, because the DPWH accounts for a very high percentage of the nation's total climate budget. Under the Local Government Act of 1991, which is supposed to operate consistently with the Climate Change Act of 2009, some public roads and infrastructure projects have been devolved to local governments. However, the ability of an LGU to obtain funds from national road network projects that are not subject to this transfer of authority is expected to determine their CCA investment ratio to some extent. Thus, it can be inferred that when the national government and joint contributions constitute the main funding sources for CCA programs, the amounts received are greater than if the funding had come from other sources, which results in higher CCA investment amounts and ratios. Therefore, CCA investment distribution is likely closely related to the relationship between central and local governments.

Additionally, the extent to which specific PPAs account for adaptation measures in each LGU is related to CCA investment amounts and ratios. For example, in Bohol Province, where the disclosures of original annual investment programs have been particularly promoted and have a higher than average CCA investment ratio, the scope of adaptation measures is extensive. Bohol Province's CCA investment in its annual investment program for the 2016 fiscal year consisted of three administrative services: general public services, economic services, and social services. CCA investment in economic services consists mainly of bridge construction, road development, maintenance (paving with concrete or asphalt), and repair. For social services, dengue fever, rabies prevention and control, and hospital construction accounted for a large share of CCA investments (Provincial Government of Bohol 2016). Thus, Bohol defines "CCA programs" broadly, and categorizing such a wide range of programs as CCA investment results in its high CCA investment ratio and amount. It is possible that such extensive climate adaptation measures have also been observed in other provinces.

5.5 Discussion

In attempting to understand the current circumstances of CCA, the validity of categorizing projects as CCA measures could be questioned. For example, in the Bohol provincial government's annual investment program, the "expected outputs" of road paving or rehabilitation projects that were considered CCA were primarily "easy access for the transport of people, crops and goods" (Provincial Government of Bohol

2016). These projects appear outside the framework of climate-resilient infrastructure, as per the typologies of PPAs for CCA designated by the local CCET process. They were probably relabeled as CCAs due to the pressure to promote CCA at the subnational level and are distinguished from "concrete policies" that are intentionally designed to adapt to long-term and additional climate change impacts (Dupuis and Biesbroek 2013). At the same time, even though specific PPAs do not address definite climate hazards based on the "impact focus" approach (McGray et al. 2007; Tanner and Horn-Phathanothai 2014), they might be considered part of a "vulnerability focus" approach. This approach includes conventional development actions, such as improving health or poverty alleviation (McGray et al. 2007; Tanner and Horn-Phathanothai 2014), given the discussion on the adaptation-development continuum (McGray et al. 2007; Tanner and Horn-Phathanothai 2014). If specific PPAs can be regarded as either of the two approaches or fall somewhere in the middle, the inclusion of specific PPAs in the CCA category can be considered valid. Because different interpretations of adaptation matter in the context of policymaking (Dupuis and Biesbroek 2013; Purdon and Thornton 2019), the broadness of CCA concepts creates space for including many types of actions. Thus, their validity needs to be considered carefully.

Regarding the relationship between CCA investment and flood risk, the findings differ from the assumptions of the research mentioned in the literature review. While prior research led to the expectation that flood risk would be related to allocations of CCA investment (e.g., Musah-Surugu et al. 2018; Karim and Noy 2020), this study could not attribute the differences between LGU CCA investment and flood risk. Therefore, the two do not appear to be clearly linked.

Concerning the relationship between fiscal scale and CCA investment, the results of this study can be contrasted with those of previous environmental expenditure and national level studies. As discussed in the literature review, prior research has indicated that population and its accompanying fiscal scale are determinants of environmental expenditures. However, the findings of this study do not show any relationship between these variables and the CCA investment ratio. Ishiwatari (2019) conducted a national level comparative study of Asian countries, including the Philippines. The findings indicated that economic growth leads to greater fiscal leeway, in turn leading to increased investment in flood prevention. In contrast, in the present study, the local level fiscal scale was not correlated with the CCA investment ratio, but rather with the CCA investment amount. Accordingly, at the subnational level, the relationship between financial scale and CCA investment depends on whether CCA investment is measured in terms of ratios or investment amounts. Thus, the manner in which CCA investment is measured should be carefully considered when investigating the relationship between fiscal scale and CCA investment.

In addition to the factors mentioned in the previous section, the likelihood of political connections, including nepotism, increases the CCA investment ratio of some LGUs and includes behaviors such as giving preferential treatment to the hometowns of key members of the CCC. The "eco-town" is a key CCC's policy. The first area selected for this designation was Surigao del Norte Province, which is the CCC vice-chairperson's hometown. More than 40% of the CCC's annual budget was expected to be allocated to eco-town projects (Barlaan and Christian 2011). Whether this project in Surigao del Norte was considered part of the province's 2016 CCA investment remains unclear. However, nepotism may also affect the distribution of funds.

Although not this study's direct focus, the nepotistic aspects of budget allocation can also be observed in the aforementioned PSF. Sarangani Province, the electoral district of a congress member who promoted the passage of the PSF bill and was one of the first congressmen to sign this bill as a co-author (Romero 2011), was one of the first provinces selected for the PSF. Attempts to steer financing from funds to electoral districts have also been made. For example, the finance secretary was dispatched to the province of Eastern Samar (Sunstar 2018), where the congress member representing this province appealed for support for the PSF bill (Romero 2011). On that occasion, explanations on PSF were provided, and further explanations were to be offered (Sunstar 2018). The PSF is not well-known (TDRI 2019), and knowledge of the fund's existence and opportunities to obtain detailed information could increase access to this fund. This suggests that the opportunity to receive such explanations on PSF is another example of the consideration given to congress members' home districts.

As a background to the focused allocation of funds to specific regions or considerations, Funatsu (2018) highlighted that, in the Philippines, local fiscal negotiations are left to negotiations among politicians, and the importance of access to political channels in the politics of budget acquisition is likely to be higher than that of administrative channels. Conversely, other studies have shown that political connections with central governments seem unrelated to the allocation of CCA and DRR funding to local governments (Karim and Noy 2020), although a method for measuring political connections requires careful comparison. In contrast, the findings of this study can be explained by a context specific to the Philippines. This echoes the call by Purdon and Thornton (2019) to consider political factors and circumstances specific to developing countries in adaptation policy research.

5.6 Conclusion

This study aimed to examine approaches to monitoring CCA investment by local governments by observing the current status of CCA investment by local governments in the Philippines and examining the factors that influence this investment. This study addresses the gap in existing academic research that has not focused on CCA investment by local governments or the determinants of such investment. The results indicate no clear linkage between flood risk and LGUs' CCA investment amounts or ratios. There was also no relationship between fiscal scales and CCA investment ratios. Instead, differences in financing sources for CCA measures that reflect the relationships between national and local governments, the scope of CCA measures, and political connections are related to higher CCA investment.

According to this study's findings, the following three points can be made regarding CCA investment monitoring. This monitoring is assumed to be conducted by ministries responsible for budgeting and management, climate change policy formulation and monitoring, and affairs related to the local governance, as in the case of the CCET system. First, monitoring must be strengthened to ensure that CCA investment is allocated at the local level according to its hazard to the impacts of climate change. This is because allocating limited resources to CCA investments according to the relative levels of climate change risk is effective in terms of CCA. Second, as responses to the impact of climate change become increasingly necessary, both the scale and ratio of CCA investment should be monitored. The study results indicate that the slight tendency for CCA investment amounts to grow in response to LGUs' increased fiscal scale (the total amount of annual investment programs) does not mean that the CCA investment ratios will also grow. The ratio of CCA investment should also be monitored to ensure the promotion of CCA among other priority policy areas.

Lastly, the extent to which CCA investment is used for "additional" CCA measures must also be monitored. In the annual investment programs covered in this study, most PPAs tagged as CCA investments appear to have been programs and projects that had already existed. As discussed in the frequently referenced CCA continuum of approaches (McGray et al. 2007), CCA policy approaches are broad. The extent to which investment was being made in additional CCA measures that directly contribute to reducing the risk of climate change impacts also needs to be better understood by monitoring CCA investment.

Current monitoring involves collecting quantitative information through CCET. The amounts invested in CCA, climate change mitigation, related measures, and the number of sectors and types of measures tied to climate change in annual investment programs seem to be prioritized. Based on the quality review and assurance system, further verification of the validity of the linkage between individual measures (PPAs) and typologies is expected to be conducted using the annual investment program of each LGU in terms of the effectiveness of CCA. The relatively large amount of investment in hard infrastructure contributes to the overall amount of investment in CCA. Thus, CCET needs to be supplemented with greater consideration of additional contributions to CCA, while functioning as a climate-related investment monitoring system.

A limitation of this study is the lack of analysis and discussion based on local governments' original documents on annual investment programs. This limitation is partly due to inadequate disclosure. Future research should perform a more detailed analysis using original documents from each LGU.

Despite this limitation, this study contributes to the literature on CCA policy and investment in general and its neighboring DRR and environmental policies at the subnational level. The study results contrast with existing research, especially in terms of the relationship between the elements related to physical and socioeconomic vulnerabilities, financial scales, and CCA investment. The focus on subnational level domestic investment is important to understand adaptation investment as a whole, as local governments are frontline entities in planning and implementing CCA.

Acknowledgements The author is grateful to the Department of International Studies at the University of Tokyo for their support with this research. This study was partly supported by the collaborative research project of the International Research Institute of Disaster Science, Tohoku University.

References

Andreas M, Epistola RU, Cea MA et al (2018) Multi-level climate governance in the Philippines: shaping connections for climate action. https://www.adelphi.de/en/publication/multi-level-climate-governance-philippines. Accessed 20 June 2021

Bacot AH, Dawes RA (1997) State expenditures and policy outcomes in environmental program management. Policy Stud J 25(3):355–370. https://doi.org/10.1111/j.1541-0072.1997.tb00027.x

Barlaan KA, Christian (2011) An act of God. Manila Standard. https://manilastandard.net/mobile/article/101820. Accessed 20 June 2021

Barrett S (2014) Subnational climate justice? Adaptation finance distribution and climate vulnerability. World Dev 58:130–142. https://doi.org/10.1016/j.worlddev.2014.01.014

Barrett S (2015) Subnational adaptation finance allocation: comparing decentralized and devolved political institutions in Kenya. Glob Environ Pol 15(3):118–139. https://doi.org/10.1162/GLEP_a_00314

Broietti C, Flach L, Rover S et al (2018) Public expenditure and the environmental management of Brazilian municipalities: a panel data model. Int J Sustain Dev World Ecol 25(7):630–641. https://doi.org/10.1080/13504509.2018.1485599

Climate Change Commission (n.d.) Local Climate investment brief 2016: an analytical review of the local climate expenditure

Climate Change Commission (2016) Climate budget brief 2016: an analytical review of the national climate budget. https://climate.gov.ph/files/Fy%2016%20CBB.pdf. Accessed 21 June 2021

Climate Change Commission, Department of Budget and Management (2016) National climate budgeting: a reference guide to climate budgeting at the national level in the Philippines. https://climate.gov.ph/files/NCB%20Ref%20Guide.pdf. Accessed 21 June 2021

Department of Budget and Management, Climate Change Commission, Department of the Interior and Local Government (n.d.). Climate Change Expenditure Tagging for Local Government. https://climate.gov.ph/files/CCET%20LGU%20Final.pdf. Accessed 10 Oct 2021

Department of Finance (2015) LGU Annual Regular Income (ARI). https://blgf.gov.ph/lgu-fiscal-data/. Accessed 10 Oct 2021

de Leon EG, Pittock J (2017) Integrating climate change adaptation and climate-related disaster risk-reduction policy in developing countries: a case study in the Philippines. Clim Dev 9(5):471–478. https://doi.org/10.1080/17565529.2016.1174659

Dupuis J, Biesbroek R (2013) Comparing apples and oranges: the dependent variable problem in comparing and evaluating climate change adaptation policies. Glob Environ Change 23(6):1476–1487. https://doi.org/10.1016/j.gloenvcha.2013.07.022

Funatsu T (2018) Kaidai Tounan Ajia Jichitai ni okeru yosan no seiji. Tounan Ajia no jichitai sa-bei hikaku no tame no kaidai to de-ta sakusei (Review: Politics of Local Financing in Southeast Asia), 76–96 https://www.ide.go.jp/library/Japanese/Publish/Reports/InterimReport/2017/pdf/2017_1_40_013_ch06.pdf. Accessed 21 June 2021

Government of the Philippines (n.d.) People's survival Fund. http://psf.climate.gov.ph/view/?page=home. Accessed 21 June 2021

Ishiwatari M (2019) Investing in disaster risk reduction: Scale and effect of investment in flood protection in Asia. Contributing Paper to GAR 2019. https://www.undrr.org/publication/investing-disaster-risk-reduction-scale-and-effect-investment-flood-protection-asia. Accessed 21 June 2021

Islam S, Zobair KM, Chu C et al (2021) Do political economy factors influence funding allocations for disaster risk reduction? J Risk Financ Manag 14(2):85. https://doi.org/10.3390/jrfm14020085
Karim A, Noy I (2020) Risk, poverty or politics? The determinants of subnational public spending allocation for adaptive disaster risk reduction in Bangladesh. World Dev 129:104901. https://doi.org/10.1016/j.worlddev.2020.104901
McGray H, Bradley R, Hammill A (2007) Weathering the storm: Options for framing adaptation and development. https://www.wri.org/research/weathering-storm. Accessed 21 June 2021
Miller SJ, Vela MA (2014) Is disaster risk reduction spending driven by the occurrence of natural disasters? Evidence from Peru (No. IDB-WP-500). IDB Working Paper Series. Inter-American Development Bank, Peru. https://publications.iadb.org/publications/english/document/Is-Disaster-Risk-Reduction-Spending-Driven-by-the-Occurrence-of-Natural-Disasters-Evidence-from-Peru.pdf. Accessed 21 June 2021
Mines and Geosciences Bureau (n.d.) Mines and Geosciences Bureau central database: Hazard susceptibility ratings. http://databaseportal.mgb.gov.ph/#/public/hazard-susceptibility-ratings. Accessed 21 June 2021
Musah-Surugu JI, Owusu K, Yankson PWK et al (2018) Mainstreaming climate change into local governance: financing and budgetary compliance in selected local governments in Ghana. Dev Pract 28(1):65–80. https://doi.org/10.1080/09614524.2018.1398717
Oulu M (2015) Climate change governance: Emerging legal and institutional frameworks for developing countries. In: Leal Filho W (ed) Handbook of climate change adaptation. Springer, Berlin, Heidelberg, pp 227–250
Pannier E, Vu TC, Espagne E et al (2020) The three dialectics of adaptation finance in Vietnam. Sustainability 12(18):7691. https://doi.org/10.3390/su12187691
Provincial Government of Bohol (2016) Annual Investment Program (AIP) for CY2016. By Program/Project/Activity by Sector. https://ppdo.bohol.gov.ph/plan-reports/development-plans/annual-investment-program/. Accessed 21 June 2021
Philippine Statistics Authority (2015a) City and municipal-level small-area poverty estimates 2009, 2012, and 2015. https://psa.gov.ph/content/psa-releases-2015-municipal-and-city-level-poverty-estimates. Accessed 21 June 2021
Philippine Statistics Authority (2015b) Updated 2015 and 2018 full year poverty statistics. https://psa.gov.ph/poverty-press-releases/nid/162559. Accessed 21 June 2021
Philippine Statistics Authority (2015c) The 2015 census of population and housing. https://psa.gov.ph/population-and-housing/statistical-tables. Accessed 21 June 2021
Purdon M, Thornton P (2019) Research methodology for adaptation policy analysis: Embracing the eclectic messy centre. In: Keskitalo ECH, Preston BL (eds) Research handbook on climate change adaptation policy. Edward Elgar Publishing, Cheltenham, UK.
Raquedan PB (2010) Climate change and human security in the Philippines: government policies, assessments and public opinion. In: Marquina A (ed) Global warming and climate change: Prospects and policies in Asia and Europe. Palgrave Macmillan, UK, pp 444–461
Republic of the Philippines (1991) Creating an inter-agency committee on climate change. https://climatechange.denr.gov.ph/images/DL_Files/Laws%20on%20CC/AO%20171.pdf. Admin Order 171 (May 08). Accessed 20 June 2021
Romero P (2011) Tanada, Pacquiao to back People's survival fund. Philippine Star. https://www.philstar.com/headlines/2011/02/08/655118/taada-pacquiao-back-peoples-survival-fund. Accessed 21 June 2021
Sova CA, Schipper ELF (2019) Adaptation policy and planning in developing countries. In: Keskitalo ECH, Preston BL (eds) Research handbook on climate change adaptation policy. Edward Elgar Publishing, Cheltenham, UK
Sunstar (2018) DOF: Tap funds for climate change, disaster risk reduction. https://www.sunstar.com.ph/article/424490/Sports/DOF-Tap-funds-for-climate-change-disaster-risk-reduction. Accessed 21 June 2021
Tanner T, Horn-Phathanothai L (2014) Climate change and development. Routledge

Thailand Development Research Institute (2019) Analysis of the people's survival fund: insights on climate finance delivery in the Philippines. https://tdri.or.th/wp-content/uploads/2020/04/Revised_Final_Research_Report_Yanquiling_Rhomir.pdf. Accessed 21 June 2021
World Food Programme (2014) ICA Philippines, 2014: flood risk, 2014. https://geonode.wfp.org/layers/geonode%3Aphl_ica_floodrisk_geonode_mar2014. Accessed 1 July 2021

Chapter 6
Financing Infrastructure of DRR: Possibility of Integrating Investments in Maintenance and DRR

Daijiro Mizutani

Abstract In Japan, the aging of infrastructure is progressing and the infrastructure management costs are expected to increase. Furthermore, to reduce the damage caused by frequent disasters to infrastructure, policies that ensure the resilience of infrastructure to disasters, as well as infrastructure management policies, must be implemented. In this chapter, we consider a hypothetical situation in which the infrastructure deteriorates over time and the damage caused by a disaster varies according to the state of the infrastructure's deterioration. In such a case, it is thought that infrastructure management policies during normal times (i.e., not during disasters) also serve as disaster prevention policies, and a method is introduced to optimize management policies by taking both the lifecycle costs and disaster risks of infrastructures into account. Sensitivity analyses are then conducted on the probability of disaster occurrence, the rate of deterioration of infrastructure, and the relationship between the state of deterioration of infrastructure and the damage caused by disasters, to derive optimal management policies for various situations. We specifically assume that the probability of disaster occurrence increases over time, and we propose that optimal management policies during normal times may change meaningfully depending on the probability of disaster occurrence. The results are then discussed in the context of infrastructure finance.

Keywords Infrastructure asset management · Disaster risk reduction · Infrastructure finance · Optimal control theory

6.1 Introduction

It is essential, for social and economic activities, to keep infrastructure and its network always usable. Also, the risk caused by both deteriorations of infrastructure and external influence of a disaster such as an earthquake would reduce users' safety. Tra-

D. Mizutani (✉)
Department of Civil and Environmental Engineering, Tohoku University, 6-6-06, Aoba, Aramaki, Aoba, Sendai, Miyagi 980-8579, Japan
e-mail: daijiro.mizutani.a5@tohoku.ac.jp

M. Ishiwatari and D. Sasaki (eds.), *Financing Investment in Disaster Risk Reduction and Climate Change Adaptation*, Disaster Risk Reduction,
https://doi.org/10.1007/978-981-19-2924-3_6

ditionally, investment in the maintenance of deteriorating infrastructure and investment in infrastructure disaster prevention have often been regarded as two separate investment activities. In this chapter, we discuss the usefulness of integrating these two types of investments. Specifically, we demonstrate through numerical analysis that routine investment (i.e., investment in maintenance) can also be regarded as investment in disaster prevention when the risks of both infrastructure deterioration and disasters are considered simultaneously.

To minimize such risks, it is important to rationally and optimally maintain infrastructure. Furthermore, in Japan, significant infrastructure built during the period of high economic growth (from 1954 to 1973) has deteriorated. Given the difficulty in securing a budget to respond to the deterioration of such infrastructure, a certain level of management that can keep the level of risk below a certain level should be implemented to reduce management costs as much as possible. Through analyses, simulations, and other methods, the possibility that damage at the time of an earthquake could differ depending on the deterioration condition of the infrastructure (in other words, the deterioration could lower earthquake resistance) has been noted in recent years. Furthermore, as for earthquake occurrence probability, research has accumulated in the fields of science and technology for quantification such as "the probability of earthquake occurrence in the next certain years (earthquake occurrence probability) (The Headquarters for Earthquake Research Promotion, 2005) is XX%". If the damage caused by an earthquake differed depending on the deterioration state of the infrastructure, the optimal management policy under normal (daily) conditions would also differ depending on the probability of an earthquake occurring. The optimal management policy, for example, would gradually change over time by updating the earthquake occurrence probability. Alternatively, the optimal management policy would be drastically altered if the occurrence probability of a consolidated type of earthquake increased significantly as a result of a partial occurrence of a plate boundary earthquake in the Nankai Trough. In this study, by modeling the deterioration and repair process of infrastructure with the Markov process model, how the daily infrastructure management policy would change if the risk of earthquake is considered or not is investigated assuming damage at the time of earthquake is different depending on the deterioration condition of infrastructure.

6.2 Existing Studies

6.2.1 Optimal Management Policy

Since the 1990 s, various studies have explored the optimal management policy for infrastructure. Madanat (1993) addressed the optimization problem on infrastructure repair actions as the Markov decision process (Howard 1960) and proposed a solution using dynamic programming (Bellman 1954) for the Markov process model used in this study. Jido et al. (2008) applied Madanat's methodology (Madanat 1993) to a description of infrastructure deterioration in continuous state space. Studies have also been conducted to investigate the optimization of a management policy using the latent Markov decision process when a random error is included in the inspection results for infrastructure (Madanat 1991; Madanat and Ben-Akiva 1994). A study on the steady state of the Markov process model was also considered in this research (Li and Madanat 2002). A study on the further application of this model to a continuous state space and continuous time axis was completed (Ouyang and Madanat 2006). Furthermore, an unsteady Markov process model (Aoki et al. 2006) based on the time-dependent Markov transition probability of the multistage Weibull hazard model (Kobayashi et al. 2010) and a mixed Markov process model (Kobayashi et al. 2012) based on multiple unsteady Markov transition probabilities of the hierarchical hidden Markov deterioration model was developed. A methodology for determining the optimal synchronization policy for the timing of inspection and repair for composite facilities comprised of various types of facilities has also been developed (Otazawa et al. 2008). Methodologies have been developed that not only determine the lifecycle cost and the risk management index but also incorporate the Markov decision model into an accounting system for infrastructure and Fault Tree Analysis (Hori et al. 2011; Kaito et al. 2011). A model that quantifies an economic analysis of an inspection action by using Real Options Analysis together (Kobayashi et al. 2011), and an optimal model for scrapping and repair considering a policy for disposal of infrastructure have been also developed (Obama et al. 2012). Furthermore, during recent years, studies on the optimization problem at a network level considering the linkage to a network for infrastructure have also been completed (Torres-Machi et al. 2014; Lethanh et al. 2017; Burkhalter et al. 2017; Mizutani et al. 2017, 2020). However, in the aforementioned studies, a management policy for infrastructure was optimized by solving the optimization problem with the cost of management such as inspection, repair, and updating as an objective function and the management level for the risk caused by deterioration as the risk control level, or the optimization problem with the objective function defined as the total surplus consisting of the management cost, user benefit, and the risk caused by deterioration. As a result, the earthquake risk that changes following the deterioration condition of the infrastructure is not taken into account in these studies. Although there are studies that qualify the variation in the vulnerability of a network for infrastructure depending on the deterioration condition of the infrastructure (Dehghani et al. 2015) and propose a method to optimize a management policy considering the relationship between the deterioration condition and the damage caused by an earthquake (Mayet and Madanat 2002), the previous studies do not include the methodology that would optimize a management policy under normal conditions considering the relevance between the deterioration condition and the damage caused by an earthquake whose occurrence probability is increasing over time.

6.2.2 Relationship Between Deterioration Condition and Damage by Earthquake

Numerous studies have been completed evaluating the earthquake resistance of deteriorated facilities through simulations and experiments. Ibarra et al. (2012) used the finite element method to evaluate the earthquake resistance of an RC element of a nuclear facility with deterioration over many years and concluded that a concrete crack and corrosion of the reinforcement bar caused by deterioration over many years could have a significant influence on the earthquake resistance of an RC element. A study demonstrated the residual performance of a lead rubber bearing (LRB) that had deteriorated over many years (Hayashi et al. 2014. It was concluded that deformability is secured in a deteriorated LRB, but the influence of deterioration can be recognized in the horizontal shearing strain during the ultimate limit test. An analytical study (Dang et al. 2015) on the LRB conducted thereafter by the incremental dynamic analysis using many seismic motion input waves indicated that the change in failure probability of a bridge with a deteriorated bearing is smaller compared to that with a bearing that is not deteriorated in the case of a small-scale earthquake, while the failure probability of a bridge caused by deterioration of a bearing could increase in the case of an earthquake with more than 1,000 gal. Furthermore, Onodera et al. (2015) used incremental dynamic analysis to examine the impact of a seismic bearing deterioration over many years on the earthquake response of an RC bridge system with a seismic bearing. It was demonstrated that a decrease in damping performance caused by deterioration of a seismic bearing would result in a significant advancement of plasticization of an RC pier in an RC bridge system and a reduction in the strength of seismic motion. Thanapol et al. (2016) proposed a methodology for updating the earthquake resistance of RC structures in a coastal area based on the progression of corrosion of the reinforcement bar using inspection data and nondestructive inspection results. As shown in the previous studies, a difference in the damage at the time of an earthquake caused by the deterioration of infrastructure over many years cannot be completely denied. However, previous studies on the optimization of infrastructure management policies lack the methodology that would optimize a management policy under normal conditions while taking into account the relevance of the deterioration condition and the damage caused by an earthquake whose occurrence probability is increasing over time.

6.3 Model and Optimal Repair Policy

A deterioration and repair process of a bridge element is modeled as a Markov process. A discrete time axis is defined as $t = t + 1$ $(t \in \{0, 1, \dots, T - 1\})$. t indicates a time point in which inspection and repair decisions are conducted, and the length of the time interval between t and $t + 1$ is denoted by z. Tz is the length of the planning period. Supposing the deterioration condition of the bridge element is evaluated in I

condition states (CSs), $1, 2, \ldots, I$. The larger the CS is, the worse the deterioration condition is. Whether or not the bridge element is repaired at t is determined based on the inspection result of the CS at t. One type of repair is considered, and the CS is recovered to 1 when repaired. Assuming the repair can be conducted in a relatively short period, time to repair is not considered. When the CS is i at t, the discounted present value of the total cost at t is defined as

$$l_t(\boldsymbol{\delta}_t|i) = \delta_{t,i}n + u(p_i + q_i f(t)) + \rho^z \sum_{j=1}^{I} \pi_{i^{(1-\delta_{t,i})},j}\, l_{t+1}(\boldsymbol{\delta}_{t+1}|j), \tag{6.1}$$

$$l_T(\boldsymbol{\delta}_T|i) = \delta_{T,i}n + u(p_i + q_i f(T)), \tag{6.2}$$

where

- $\delta_{t,i}$: Decision variable of repair for CS i at t ($\delta_{t,i} = 1$ when a repair is conducted for CS i at t, and $\delta_{t,i} = 0$ when a repair is not conducted for CS i at t.)
- $\boldsymbol{\delta}_t$: Decision variable vector defined as $\boldsymbol{\delta}_t = \{\delta_{t,1}, \ldots, \delta_{t,I}\}$
- n: Unit repair cost
- u: Unit user cost (User cost when the bridge element becomes unavailable.)
- p_i: Probability that user cost u is needed when the CS is i, due to deterioration
- q_i: Probability that user cost u is needed when the CS is i, due to the earthquake when the earthquake occurs (Note that since q_i depends on i, the deterioration state and the probability of incurring user costs due to an earthquake are interdependent.)
- $f(t)$: Earthquake occurrence probability in $[t, t+1)$ (a monotonically increasing function with respect to t)
- ρ: Discount factor
- $\pi_{i,j}$: Markov transition probability from CS i to CS j with time interval z
- $i^{(1-\delta_{t,i})}$: CS after repair at t when CS before repair at t is i.

The optimal value function of $l_t(\boldsymbol{\delta}_t|i)$ is defined as $v_t(i)$. Assuming $\boldsymbol{\delta}_T$ is given, by backward recursion, the optimal repair policy at t, $\boldsymbol{\delta}_t^*$, can be found by

$$\boldsymbol{\delta}_t^* = \underset{\boldsymbol{\delta}_t \in \boldsymbol{\Delta}}{\arg\min}\ \delta_{t,i}n + u(p_i + q_i f(t)) + \rho^z \sum_{j=1}^{I} \pi_{i^{(1-\delta_{t,i})},j}\, v_{t+1}(j), \tag{6.3}$$

where the state space of $\boldsymbol{\delta}_t$ is denoted as $\boldsymbol{\Delta}$, which is Ith Cartesian power of (0, 1). Note that $v_T(j) = \delta_{T,i}n + u(p_i + q_i f(t))$.

6.4 Case Study

Assuming a bridge element such as a rubber bearing, the optimal management policy is derived from the proposed model considering the earthquake risk increasing over time. The input conditions are listed in Table 6.1. Those conditions are set with reference to Road Bureau of Ministry of Land, Infrastructure, Transport and Tourism (2014), Osaka Prefecture (2016), and O'Reilly et al. (1994), assuming an actual situation of bridge management in Japan.

The values in the table are hypothetical, set to imitate reality. In the actual policy analysis, it is desirable to determine the unit repair cost based on the information of the costing standard and the past repair history, and the unit user cost based on the traffic demand data, respectively. As the unit repair cost decreases or the unit user cost increases, the preventive maintenance policy will be more likely to be selected as the optimal policy. The discount factor ρ was set to the same value as in the cost-benefit analysis manual of the Ministry of Land, Infrastructure, Transport and Tourism (2018). The Markov transition probabilities to express the stochastic deterioration process of the bridge element are shown in Table 6.2, and the evolution process of the state vector following the Markov transition probabilities is illustrated in Fig. 6.1. The function of the earthquake occurrence probability with respect to t is assumed as a simple linear function $f(t) = az\lfloor t/z \rfloor$. a is a parameter. In this case study, we consider three scenarios, supposing megathrust earthquakes, as

Scenario 1 : High earthquake occurrence probability scenario ($a = 1/100$)
Scenario 2 : Low earthquake occurrence probability scenario ($a = 1/200$)
Scenario 3 : No earthquake occurrence probability scenario ($a = 0$).

Specific $f(t)$ in each scenario is graphically shown in Fig. 6.2.

In each scenario, the optimal repair policy $\delta_t^* = (\delta_{t,1}^*, \delta_{t,2}^*, \dots, \delta_{t,I}^*)$ ($t \in \{0, 5, 10, \cdots, 100\}$) are derived. At all time points in every scenario, $\delta_{t,i_1}^* \leq \delta_{t,i_2}^*$ was satisfied when $i_1 < i_2$. Here, we define CS_t^* as

Table 6.1 Input conditions

Description	Variable	Value
Number of CSs	I	4
Number of time points	T	20
Inspection interval	z	5
Unit repair cost	n	940
Probability (deterioration)	p_1	0
	p_2	0
	p_3	0
	p_4	0.1
Probability (earthquake)	q_1	0
	q_2	0
	q_3	0
	q_4	0.5
Unit user cost	u	65,000
Discount factor	ρ	0.96

Table 6.2 Markov transition probability $\pi_{i,j}$

$\downarrow i, \rightarrow j$	1	2	3	4
1	0.9800	0.0150	0.0040	0.0010
2	0	0.9870	0.0100	0.0030
3	0	0	0.9950	0.0050
4	0	0	0	1

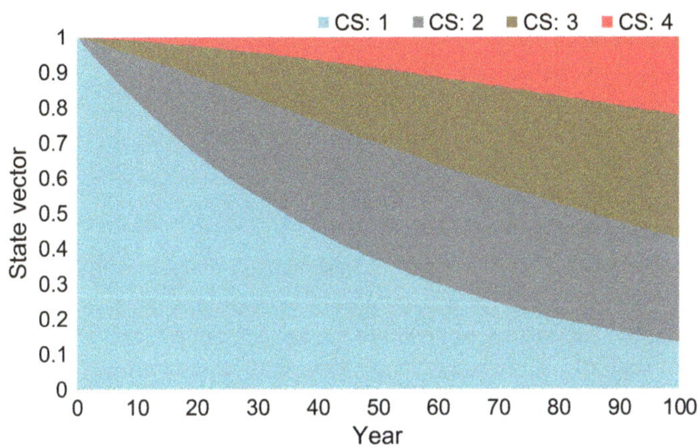

Fig. 6.1 Deterioration process

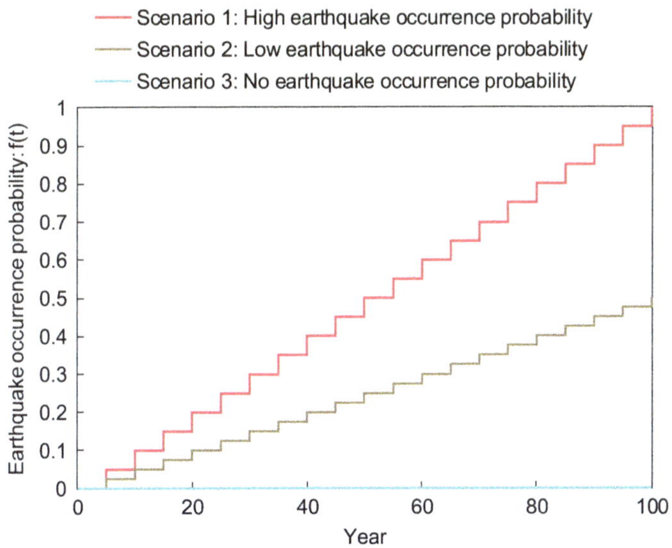

Fig. 6.2 Earthquake occurrence probability

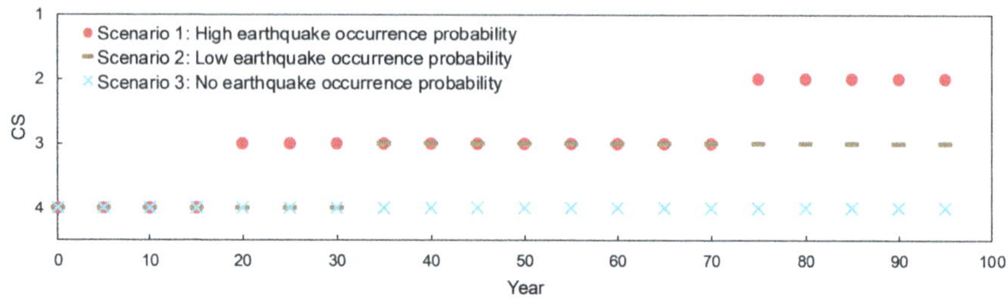

Fig. 6.3 Optimal repair policies

$$CS_t^* = \min i \;\; \text{s.t.}\; \delta_{t,i}^* = 1. \tag{6.4}$$

CS_t^* in each scenario is shown in Fig. 6.3. In scenario 1 and scenario 2 in which earthquake occurrence probabilities are increasing over time, CS_t^* is becoming smaller over time, i.e., the superiority of preventive management policies is increasing over time. CS_t^* in scenario 3 is always larger than that in scenario 2 or equal to it, as the earthquake occurrence probability in scenario 3 is higher than that in scenario 2. On the other hand, in scenario 3 whose earthquake occurrence probability is 0 constantly, CS_t^* does not change from 4 over time and the corrective management policy is superior at any time.

In this case study, we created scenarios based on realistic situations, and the results show that the optimal management policy may change depending on the probability of earthquake occurrence and its evolution over time. This implies that the risk of an earthquake over time can be reduced by changing the management policy when the current management policy is determined excluding the risk of an earthquake.

6.5 Implications from a Financing Perspective

This study shows that, in the problem of optimizing the maintenance policy of infrastructures, the optimal management policy can be different when the probability of disaster occurrence varies (i.e., the optimal repair policies are different for each scenario in Fig. 6.3), under the assumption that the damage caused by disasters differs depending on the state of deterioration. For instance, suppose the optimal policy is changed from a corrective maintenance policy to a preventive maintenance policy by taking disaster risk into account in a situation where the optimal policy is a corrective maintenance policy that only minimizes maintenance costs without taking disaster risk into account (i.e., a preventive maintenance policy has higher maintenance costs than a corrective maintenance policy). This means that the cost difference between the corrective and preventive maintenance policies is invested in disaster prevention. In other words, it implies that disaster reduction investment can be made not only through independent disaster reduction investment, such as updating infrastructure performance solely to improve disaster reduction performance but also through routine or periodic maintenance actions, such as changes in maintenance policies.

Since interventions in infrastructure (e.g., repair, reinforcement, renewal, and improvement of disaster prevention performance) generally require fixed costs, human resources, and procurement labor per intervention, in this context, unifying or synchronizing investment in infrastructure through routine maintenance and investment in disaster prevention can bring benefits through economies of scale. To take advantage of these economies of scale, it is critical that the relationship between the state of deterioration and the damage caused by disasters be quantified in infrastructure asset management, and based on the results, a framework for minimizing the total cost of infrastructure be developed, in which the cost of disaster damage is internalized in addition to the cost of maintenance. Furthermore, quantifying future disaster risks and intervening in infrastructure based on these risks are desirable. Synchronization of the two interventions, instead of independent interventions for disaster management and maintenance, may lead to cost savings and more efficient infrastructure DRR financing through economies of scale.

In the area of infrastructure management in Japan, PPPs and concession schemes have been introduced to make use of private sector funds and technologies. Furthermore, the application of the optimal maintenance policy derived in this study based on the Markov decision process corresponds to the implementation of condition monitoring maintenance. As a result, it is also critical to develop management accounting systems (e.g., renewal accounting, and deferred maintenance accounting) to fund the implementation of infrastructure interventions that combine maintenance and disaster management. Taking these institutional designs into account, it is desired that a method of infrastructure asset management will be developed that integrates the maintenance of infrastructure and the reduction of damage in the event of a disaster.

Acknowledgements This study was partially supported by the Ensemble Project for Early Career Researchers at Tohoku University.

References

Aoki K, Yamamoto K, Kobayashi K (2006) An optimal inspection/rehabilitation model of multi-components systems with time-dependent deterioration processes. J Jpn Soc Civ Eng Ser F 62(2):240–257 (In Japanese)

Bellman R (1954) The theory of dynamic programming. Bull Am Math Soc 60(6):503–515

Burkhalter M, Martani C, Adey BT (2017) Determination of risk-reducing intervention programs for railway lines and the significance of simplifications. J Infrastruct Syst 24(1):04017038

Dang J, Higashide T, Igarashi A, Adachi Y, Hayashi T (2015) Dynamic analysis to investigate the effect of aging deterioration of lead rubber bearings on the seismic performance of bridges. J Jpn Soc Civ Eng Ser A1 71(4):I_713-I_724 (In Japanese)

Dehghani MS, Flintsch G, McNeil S (2015) Impact of road conditions and disruption uncertainties on network vulnerability. J Infrastruct Syst 20(3):04014015

Hayashi K, Adachi Y, Komoto K, Yatsumoto H, Igarashi A, Dang J, Higashide T (2014) Experimental verification for remaining performance of lead rubber bearings with aging deterioration. J Jpn Soc Civ Eng Ser A1 70(4):I_1032-I_1042 (In Japanese)
Hori M, Tsuruta T, Kaito K, Kobayashi K (2011) Maintenance management accounting system of waste water disposal systems. J Jpn Soc Civ Eng Ser F4 67(1):33–52 (In Japanese)
Howard RA (1960) Dynamic programming and Markov processes. The MIT Press
Ibarra L, Dasgupta B, Chiang K-T (2012) Seismic performance of degraded shear walls for long-term compliance periods. J Disaster Res 7(5):638–644
Jido M, Otazawa T, Kobayashi K (2008) Optimal repair and inspection rules under uncertainty. J Infrastruct Syst 14(2):150–158
Kaito K, Kanaji H, Kobayashi H, Mashima N, Ohishi H, Matsuoka K (2011) Optimum inspection policy for long span bridge based on fault tree analysis with visual inspection data. J Jpn Soc Civ Eng Ser F4 67(2):74–91 (In Japanese)
Kobayashi K, Kiyoyuki K, Lethanh N (2010) Deterioration forecasting model with multistage Weibull hazard functions. J Infrastruct Syst 16(4):282–291
Kobayashi K, Eguchi M, Oi A, Aoki K, Kaito K (2011) The optimal implementation policy of pavement inspection with deterioration uncertainty. J Jpn Soc Civ Eng Ser E1 67(2):75–90 (In Japanese)
Kobayashi K, Eguchi M, Oi A, Aoki K, Kaito K, Matsumura Y (2012) The optimal repair and replacement model of pavement structure. J Jpn Soc Civ Eng Ser E1 68(2):54–68 (In Japanese)
Lethanh N, Adey BT, Burkhalter M (2017) Determining an optimal set of work zones on large infrastructure networks in a GIS framework. J Infrastruct Syst 24(1):04017048
Li Y, Madanat S (2002) A steady-state solution for the optimal pavement resurfacing problem. Transp Res Part A: Policy Pract 36(6):525–535
Madanat S (1991) Optimizing sequential decisions under measurement and forecasting uncertainty: Application to infrastructure inspection, maintenance and rehabilitation. Doctoral dissertation, Massachusetts Institute of Technology
Madanat S (1993) Incorporating inspection decisions in pavement management. Transp Res Part B: Methodol 27(6):425–438
Madanat S, Ben-Akiva M (1994) Optimal inspection and repair policies for infrastructure facilities. Transp Sci 28(1):55–62
Mayet J, Madanat S (2002) Incorporation of seismic considerations in bridge management systems. Comput-Aided Civ Infrastruct Eng 17(3):185–193
Mizutani D, Burkhalter M, Adey BT, Martani C, Ramdas V (2017) Initial investigations into the use of three heuristic algorithms to determine optimal intervention programs for multiple railway objects. Int J Arch Eng Constr 6(3):1–20
Mizutani D, Nakazato Y, Lee J (2020) Network-level synchronized pavement repair and work zone policies: optimal solution and rule-based approximation. Transp Res Part C: Emerg Technol 120:102797
Obama K, Kaito K, Aoki K, Kobayashi K, Fukuda T (2012) The optimal scrapping and maintenance model of infrastructure considering deterioration process. J Jpn Soc Civ Eng Ser F4 68(3):141–156 (In Japanese)
Onodera M, Matsuzaki H, Suzuki M (2015) Effect of deteriorated isolators on the seismic response of reinforced concrete columns with isolator. J Jpn Soc Civ Eng Ser A1 71(4):I_737-I_748 (In Japanese)
O'Reilly D, Hopkin J, Loomes G, Jones-Lee M, Philips P, McMahon K, Ives D, Soby B, Ball D, Kemp R (1994) The value of road safety: UK research on the valuation of preventing non-fatal injuries. J Transp Econ Policy 28(1):45–59
Osaka Prefecture (2016) Guideline for bridge inspection (In Japanese)
Otazawa T, Yamamoto K, Aoki K, Kobayashi K (2008) Optimal synchronizing repair model of highway equipment facilities. J Jpn Soc Civil Eng Ser F 64(2):200–217 (In Japanese)
Ouyang Y, Madanat S (2006) An analytical solution for the finite-horizon pavement resurfacing planning problem. Transp Res Part B: Methodol 40(9):767–778

Road Bureau and City Bureau of Ministry of Land, Infrastructure, Transport and Tourism (2018) Cost-benefit analysis manual (In Japanese)

Road Bureau of Ministry of Land, Infrastructure, Transport and Tourism (2014) Guideline for Periodic Road Bridge Inspection (In Japanese)

Thanapol Y, Akiyama M, Frangopol DM (2016) Updating the seismic reliability of existing RC structures in a marine environment by incorporating the spatial steel corrosion distribution: Application to bridge piers. J Bridg Eng 21(7):04016031

Torres-Machi C, Chamorro A, Videla C, Pellicer E, Yepes V (2014) An iterative approach for the optimization of pavement maintenance management at the network level. Sci World J Article ID 524329 (2014)

Chapter 7
Financing Climate-Resilient Coasts: Tracking Multilateral Aid for Ocean and Coastal Adaptation to Climate Change in Asia-Pacific

Nagisa Shiiba

Abstract The Asia–Pacific coastal region is at the forefront of climate change. Coastal and island communities endure climate-related hazards and marine and coastal ecosystem degradation, which repeatedly challenge their transition to a sustainable blue economy. Therefore, climate change adaptation in coastal zone communities is imperative, particularly, for developing countries where financial access is an ever-present challenge. Some existing studies have tracked global adaptation finance flows but the scale of finance in the ocean and coastal zones remains unknown. Detailed public finance monitoring would offer greater insights to close the finance gap for vulnerable coastal and island countries. This study aims to comprehend the Asia–Pacific adaptation finance landscape and show its aid flow trends and features, with a special focus on ocean and coastal areas. Results show that the amount of adaptation funds directed to the oceans and coastal regions in the Asia–Pacific has gradually increased, with a fivefold growth over the past decade. Recently, "blue finance" or ocean sustainability funding has attracted attention and accelerated concrete actions by the international community as a whole. Thus, this study also provides insight on how emerging blue finance initiatives promote climate change adaptation in the ocean and coastal areas.

Keywords Asia–Pacific · Adaptation finance · Blue finance · Climate change adaptation · Ocean and coastal zones

7.1 Introduction

Livelihoods in low-lying coastal areas and small islands are highly dependent on coastal and marine ecosystems. As climate change escalates, ocean and coastal environments change causing various hazards, including rising sea-levels, storm surges, and biodiversity degradation. The Intergovernmental Panel on Climate Change

N. Shiiba (✉)
Institute for Global Environmental Studies, Kanagawa, Japan
e-mail: shiiba@iges.or.jp

M. Ishiwatari and D. Sasaki (eds.), *Financing Investment in Disaster Risk Reduction and Climate Change Adaptation*, Disaster Risk Reduction,
https://doi.org/10.1007/978-981-19-2924-3_7

(IPCC) Special Report on Ocean and Cryosphere (SROCC) 2019 indicates that the world's oceans are transitioning to unprecedented conditions with increased temperatures, upper ocean stratification and acidification, oxygen decline, and altered net primary production (IPCC 2019). Asia–Pacific, with its myriad islands and coastal areas, is amongst the regions most vulnerable to the impacts of climate change. According to the World Risk Report 2020, which analyzes natural disasters and social vulnerability in 171 countries and assesses their disaster risk, six of the top 10 countries with the highest disaster risk (Vanuatu, Tonga, Solomon Islands, Brunei, Papua New Guinea, and the Philippines) are in the Asia–Pacific (Behlert et al. 2021). Small island developing states (SIDS) scattered across the Pacific are particularly defenseless against coastal disasters such as floods and storm surges. Therefore, investing in ocean and coastal adaptation measures in response to climate change is imperative to build resilient communities in the Asia–Pacific region.

Through climate resilience and adaptation efforts, the international community must come forward to ensure that adequate resources are mobilized to vulnerable coastal populations to help them cope with the adverse impacts of climate change. Although global climate finance flows are increasing at a fast pace, there is a significant gap in adaptation finance (Oliver et al. 2018; UN Environment Programme 2021b, a). The Organization for Economic Cooperation and Development (OECD) reported that adaptation received only one-fourth of the climate finance mobilized by developed countries in 2019 (OECD 2021). In addition to the need to strengthen adaptation funding globally, there is also a growing recognition of the challenge to ensure that such funds are appropriately allocated to vulnerable communities (Saghir et al. 2021). However, information on the allocation of funds remains scarce, and none of the analyzes have focused specifically on the ocean and coastal zones. Because climate change adaptation involves a wide range of sectors and targets, it is important to understand where adaptation funds are flowing to or not, to direct the future investment flow to strengthen ocean and coastal zones.

The investment momentum of the Sustainable Development Goal (SDG) Target 14, or the sustainable use of ocean and marine resources continues. The OECD report "The Ocean Economy in 2030" suggests that the market size of the ocean economy may reach more than US$ 3 trillion and approximately 40 million full-time employed people by 2030 (OECD 2016). "*Blue finance*" is an emerging concept that has been garnering interest among stakeholders including low-lying coastal and island countries, international donors, non-governmental development organizations, and private sectors who are engaged in ocean-related industries. In 2018, "Sustainable Blue Economy Finance Principles" were developed by the European Commission, World Wide Fund for Nature, the World Resources Institute and the European Investment Bank to accelerate activities for the SDG 14. By providing ocean-oriented investment principles for the financial community, SDG 14 aims to ensure long-term ocean-related investment gains without compromising the conservation and sustainable usage of ocean resources. With the recent rise in interest in marine and coastal issues, such as marine plastic debris, development donors are now taking a more active role in sustainable ocean finance.

This momentum is critical from the perspective of climate change because the impacts of climate change and ocean sustainability are inseparable. The IPCC SROCC projects that the impact of climate-induced ocean health deterioration on the global economy will reach US$ 428 billion per year by 2050. Coping strategies for climate risks in the ocean and coastal zones are substantial from an economic perspective. In parallel, there are growing expectations for investment in ocean-based solutions to climate change (e.g., blue carbon ecosystems, marine renewable energy). This implies that the international community, which has long considered the ocean as a victim of climate change, has reversed its thinking and intends to use oceans to solve climate change issues.

Nevertheless, such increased attention on investment in ocean-based solutions is not necessarily directed toward climate change adaptation. The international debate on oceans and climate change at the UN Climate Conference (UNFCCC COPs) focuses on the use of oceans as a mitigation measure, such as reducing emissions from the shipping sector, promoting marine renewable energy, and preserving oceans and marine ecosystems as sinks (ocean carbon ecosystems). Regarding the adaptation and ocean linkage, the Nairobi Work Programme (NWP), which aims to develop and disseminate information and knowledge on adaptation, has recently identified "oceans and coastal zones" as one of its priority issues. Under the NWP, the promotion of ocean and coastal adaptation measures began with the establishment of an expert group to compile knowledge and share information. Thus, a critical step in promoting adaptation finance is to incorporate an adaptation perspective in the ocean sustainability investment drive.

Despite a significant awareness of the importance of adaptation investments, there is a lack of understanding of the specific types of investments required for adaptation in coastal communities. This may be attributed to limited research: there are some sector-specific studies but very few studies provide an overview of ocean and coastal adaptation. Given that the use of blue finance for adaptation may serve as a new source of finance to complement traditional adaptation funding, it is important for international donors and other stakeholders to recognize the role they play in marine and coastal adaptation so that their involvement in ocean and coastal issues can be directed to adaptation. At present, there is limited existing research that provides insights into the status and challenges of aid for ocean and coastal adaptation. Furthermore, there is still a lack of information on the overall scale of funding and specific investments.

To fill these gaps, this study attempts to provide an overall picture including the status and challenges of the ocean and coastal adaptation aid projects that have been implemented in the Asia–Pacific region in the last decade. While adaptation finance is a broad concept that includes both public and private finance, the author discusses this topic specifically in the context of development aid. This study aims to investigate climate adaptation investment options in low-lying and coastal communities and understand the current state of financial mobilization for ocean and coastal adaptation in the Asia–Pacific region. The author analyzes development aid projects to understand public financial flows for ocean and coastal adaptation. The analysis employs project-level data from four important multilateral development finance institutions in the Asia–Pacific region. Specifically, the study attempts to explain the scale and the overall distribution of funds and categorize the sectors to which these funds have been mobilized. This will provide a landscape of marine and coastal adaptation aid, as well as suggestions on how the expected multilateral finance should be used to build a resilient development pathway in Asia–Pacific regions.

7.2 Blue Finance to Climate Change Adaptation

7.2.1 Overview of Blue Finance

"*Blue finance*" as a concept has no internationally agreed definition and is derived from "blue economy," which indicates the sustainable use of ocean resources for economic growth, improved livelihoods, employment, and ocean ecosystem health (World Bank and UNDESA 2017). The definitions of the blue economy and "ocean economy," which originated at the 2012 United Nations Conference on Sustainable Development held in Rio de Janeiro, may vary depending on the institution (Smith-Godfrey 2016). At present, the term, blue economy, is often used interchangeably with sustainable ocean economy, and both indicate financing efforts to direct investments and actions that promote the sustainable use of marine resources and ocean conservation. The term blue finance is used in the present study to emphasize the focus on the financing of the sustainable use and conservation of the oceans, rather than on ocean-related economic activities in general. It may be noted that there is no clear distinction between blue finance and ocean finance but some existing literature use the latter term to mean financial instruments that support sustainable development of the ocean economy (Sumaila et al. 2021).

Concrete ideas, which attract global attention for blue finance, are beginning to emerge, such as the Seychelles government's issuance of the world's first "Blue Bond" in 2018. Multilateral gold banks have also announced blue finance initiatives, such as PROBLUE, a multi-donor trust fund established by the World Bank Group in 2018. In 2019, the Asia Development Bank (ADB) committed to invest US$ 5 billion in ocean restoration and conservation in the Asia–Pacific region in its *Action Plan for Healthy Oceans and Sustainable Blue Economies*. More recently, in June 2021, the ADB and the European Investment Bank (EIB) jointly agreed on a new *clean and sustainable ocean partnership* to support the blue economy in Asia–Pacific.

Remarkably, there have been recent attempts to facilitate stakeholder engagement in blue finance by providing toolkits and concrete steps to invest in ocean sustainability. The Blue Paper published by the High-Level Panel for a Sustainable Ocean Economy proposes opportunities for actions including setting up a new shared guideline, strengthening knowledge and data, exploring new financing mechanisms, and boosting new approaches to insurance (Sumaila et al. 2020). The Friends of Ocean Action, an informal and multi-stakeholder coalition, published the Ocean Finance

Handbook (De Vos and Hart 2020). Furthermore, the UN Environment Programme recently developed a practical guide for financial institutions seeking investment opportunities in a sustainable ocean economy (UN Environment Programme 2021b, a).

7.2.2 *Climate Change and Blue Financing*

The ocean has been attracting attention in recent climate change discussions. There has been growing momentum for ocean and climate action, as illustrated by the 25th UNFCCC Conference of Parties (COP25) in 2019, named "Blue COP". The debate around climate and blue economy relationships was also significantly driven by the 2019 IPCC SROCC and the 2019 High-Level Ocean Panel's Blue Paper on climate impacts on the ocean economy (Sumaila et al. 2021). The Blue Paper, *The Expected Impacts of Climate Change on the Ocean Economy*, indicates that climate change is projected to cause significant economic damage, such as the loss of coral reef cover and reduced tourism revenues, particularly in SIDS (Gaines et al. 2019).

The ocean is a fundamentally significant part of the earth's climate system and the previous climate change debate has focused on ocean degradation due to climate change impacts. However, recent studies have highlighted the role of the ocean as a solution to climate change (Gattuso et al. 2018; Cooley et al. 2019; Hoegh-Guldberg et al. 2019a). Ocean-based mitigation options include renewable energy, transport, food production, ecosystems, and potential future contributions from carbon storage (Hoegh-Guldberg et al. 2019b). Marine and coastal ecosystems, such as coral reefs, coastal wetlands, salt marshes, and mangroves serve as adaptation measures in response to climate change (see Spalding et al. 2014 for instance). However, blue finance discourse is mostly concentrated on how to generate profits and less on how to avoid losses. Hence, ocean-based adaptation measures seem to lag behind the mitigation measures.

7.2.3 *The Role of International Aid in Ocean and Coastal Adaptation Actions*

Oceans and coastal activities involve a large variety of stakeholders. To boost the blue economy, a funding mechanism that incorporates a variety of players, including the private sector, is necessary. For climate change mitigation in general, participation from the private sector is relatively advanced through the use of green bonds and market-based mechanisms. This is becoming a reality in some sectors, such as renewable energy and sustainable transportation.

With respect to adaptation finance, however, private sector engagement remains weak, and the public sector plays a significant role because of its function (Pauw

2015). Therefore, development finance institutions and global climate funds must take the lead and influence the private sector in ocean and coastal adaptation. Existing studies have also noted the importance of achieving the inclusive blue economy governance and the role of the public sector in establishing good practices and models. In addition, international efforts through bilateral or multilateral cooperation are expected to drive effective capacity building in ocean finance (Sumaila et al. 2020). In these respects, international aid may have great significance in accelerating the adaptation of the ocean and coastal areas.

Multilateral funding plays a key role in meeting the global climate goals of the Paris Agreement. The 2020 Joint Report on Multilateral Development Banks' Climate Finance showed that climate finance mobilized by major multilateral development banks (MDBs) to low- and middle-income economies rose to US$ 38 billion (MDBs 2021). Above all, public climate finance with a specific focus on adaptation is expected to be strengthened. In 2019, the MDBs announced an increase in adaptation finance to US$ 18 billion in their climate finance targets for 2025 (MDBs 2019). It remains to be seen whether MDBs will be able to utilize this adaptation funding for the benefit of vulnerable coastal communities in the future.

7.3 Climate Change Adaptation of the Ocean and Coastal Zones in Asia–Pacific

The manifold and overreaching impacts of climate change on society and its risks to various industries and all aspects of human life must be considered. For ocean and coastal adaptation, coping strategies need to be adopted in various sectors, such as fisheries and ports. Existing literature tends to focus on adaptation measures in specific climate impacts, such as ocean acidification, but there is a lack of organized information to provide an overview of adaptation measures in the ocean and coastal zones. Therefore, in this section, the study will categorize ocean and coastal adaptation in response to climate change. Based on the literature review, the impacts of climate change and adaptation measures on ocean and coastal zones are summarized using four sectors: fisheries and aquaculture, urban infrastructure, marine and coastal ecosystems, and community disaster risk management.

7.3.1 Fisheries and Aquaculture

Marine fisheries and aquaculture largely contribute to the economies of the Asia–Pacific countries. Asia and Oceania accounted for 85.2% of the global population engaged in the fisheries and aquaculture sectors in 2016 (Food and Agriculture Organizations 2018). Small-scale fisheries and traditional fishing practices play a substantial role in the livelihood and food security of coastal communities (Lauria et al. 2018).

However, climate change has increased the pressures on fisheries and aquaculture, especially small-scale practices in the Asia–Pacific region. Lam et al. (2020) review existing studies on the impacts of climate change on tropical fisheries, including the physical and biogeochemical properties of the ocean (including warming, sea-level rise, deoxygenation, acidification, and altered nutrient concentrations), and changes in net primary production. The potential fish production across some important areas of the tropical shelf and upwelling seas in the eastern Indo-Pacific is projected to decline between 30 and 60% (Blanchard et al. 2012). Moreover, ocean acidification is expected to affect the tuna resources, which are critical to the economic development of many Pacific islands (Johnson et al. 2015).

Understanding human dimensions, including socioeconomic and cultural aspects of fishing activities, is key to build the sustainable governance of community-based resource management (Kittinger 2013). Bell et al. (2018) identify key community-based adaptation strategies in the Pacific, including vegetation management and restoration in catchments, conservation of fish habitats, landward migration of fish habitats, recruitment and production of demersal fish, efficient spatial management, and diversification of fishing methods. In addition, diverse forms of indigenous knowledge are directly or indirectly associated with climate change adaptation, as shown in a survey in India (Inaotombi and Mahanta 2018). In addition, the ecosystem approach to fisheries management helps developing countries in the Asia–Pacific region not only build resilience to climate change but also cope with the habitat degradation and overfishing which reduce the productivity of coastal fisheries (Heenan et al. 2015).

7.3.2 Urban Infrastructure

In low-lying coastal and island countries in the Asia–Pacific region, the construction and management of important infrastructures such as roads, drainage, and sewage systems are challenging and vulnerable to climate change due to Weak local governance and municipal management, high poverty incidence, and remote locations (Rahman and Rahman 2015). For instance, substantial seaport infrastructure is highly affected by extreme weather events and slow-onset events induced by climate change. Extreme weather events, including storm surges and river floods, increased tropical storm intensities, and potential changes in the wave regime damage ports and cause operational delays (Becker et al. 2018). Slow onset events, represented by sea-level rise damage to breakwaters, cause degradation of logistic functions related to marine transportation.

The coping strategies to manage the vulnerability of urban infrastructure in coastal cities include the review of land use and community development plans based on disaster risks, technological development and maintenance, monitoring of sea conditions, strategic development of coastal protection facilities, and evaluation of disaster risks (Rahman and Rahman 2015). Lam et al. (2017), in their review of the East Asian sea-ports cases, concluded that technical assessments through the modeling of climate change impacts and vulnerability mapping are substantial tools that inform decision-makers on infrastructure planning.

7.3.3 Marine and Coastal Ecosystems Service

Asia–Pacific not only comprises of rich marine and coastal ecosystems, such as coral reefs, mangroves, seagrass beds, and tidal flats, but also include shallow areas due to rising sea-levels. For example, Asia accounts for 41.9% of the world's mangrove coverage (Ward et al. 2016). Marine and coastal ecosystems, such as coral reefs, are integral for coastal tourism in many Asia–Pacific countries, which are tourism-based economies (Dearden and Manopawitr 2011). There have been past attempts to incorporate the climate change adaptation concept in the tourism sector, for instance, in Samoa and Thailand. (Wong et al. 2013; Nara et al. 2014).

Marine protected areas (MPAs) are one of the most common and representative approaches for ecosystem conservation. MPAs are used to regulate marine areas, where biodiversity conservation is particularly important, by protecting areas in advance and promoting natural regeneration efforts to restore damaged ecosystems. The establishment of MPAs is also considered to be effective for the appropriate conservation of ecosystems that are particularly vulnerable to the effects of climate change. Efforts to incorporate climate change adaptation in MPAs have already begun.

Marine and coastal ecosystems serve as adaptation measures in response to climate change impacts on human livelihoods. Nature-based solutions (NbS) are an increasingly emphasized concept for "actions for the conservation, sustainable management and restoration of natural and anthropogenically modified ecosystems that effectively and adaptively address social challenges and simultaneously provide for human well-being and biodiversity benefits" (Cohen-Shacham et al. 2016). Similarly, the use of ecosystem services as part of an overall adaptation strategy for climate change is referred to as "Ecosystem-based adaptation (EbA)." For example, the conservation and restoration of tidal flats with purification functions are expected to contribute to the purification of water in coastal areas. EbA also offers green infrastructure functions, which are often called Ecosystem-based Disaster Risk Reduction (Eco-DRR). For example, healthy coastal habitats (e.g., mangroves, sand dunes, and saltmarshes) shield coastal communities against extreme weather events such as storm surges (Agardy et al. 2017).

7.3.4 Community Disaster Management

Over the past few years, Asia–Pacific has earned global attention due to the frequent occurrence of severe mereological disasters caused due to climate change. In 2020, the largest cyclone in the history of the Bay of Bengal, "Anfang," hit India and

Bangladesh and in the South Pacific; and two powerful cyclones, "Haroldo" and "Yasa," caused human loss and material damage in Fiji and other small island countries. In the same year, the Philippines was also hit by two typhoons, which caused massive flooding in Metro Manila.

Community-based disaster management enables those countries which are front of disasters to become major participants in the implementation of disaster plans and decisions (Pandey and Okazaki 2005). Providing programs or training that increase local communities' awareness and coping capacity against natural disasters, reviewing land use planning, the advancement of weather forecasting and warning, and formulation of disaster prevention and evacuation plans are key strategies to enhance community resilience. Facilitating community-based disaster management should be supported by national policies and plans. For example, in Bangladesh, the government has developed and implemented numerous national policies with the aim of strengthening risk reduction measures at the community level by decentralizing disaster management and developing national-local partnerships (Azad et al. 2019).

7.3.5 Cross-Sectoral Approaches for Integrated Coastal Management

Policy measures that enable broader sectoral participation in marine and coastal management, such as integrated coastal zone management (ICZM) and marine spatial planning (MSP) are created to ensure environmental integrity in both development and adaptation planning (Spalding et al. 2014). ICZM, first proposed at the 1992 Earth Summit under the Agenda 21 action plan, has been widely used internationally as a tool for bringing together stakeholders in coastal areas to achieve effective and integrated coastal ecosystem management. Climate change adaptation in the coastal zone may be achieved by planning for the long-term impacts and costs of climate change in the ICZM policy cycle. The development of climate change adaptation strategies in ICZM target areas is beneficial for various stakeholders participating in ICZM and its accumulated local risk data is utilized for adaptation measures at the regional level (O'Mahony et al. 2020).

MSP has also been noted as a means to enable consensus-building in consultation with various stakeholders for the future development, use, and conservation of marine areas. EbA methods in MSP and other coastal management practices have been explored in some studies (Khan and Amelie 2015). From the Asia–Pacific perspective, Giffin et al. (2020) reviewed the cases of EbA and MSP integration in Fiji, Vietnam, Papua New Guinea, and Solomon Islands, and found that they are characterized by participatory approaches, community-based, co-management, or shared governance, and with local and regional scales of implementation.

7.4 Tracking Aid Projects for Ocean and Coastal Adaptation in the Asia–Pacific

7.4.1 Methodology

As stated in this study, there are a wide variety of cross-sectoral adaptation strategies in the ocean and coastal zones and it is essential to first understand the current funding flows to efficiently direct additional financing efforts. This study is extended to comprehend the aid flow to the ocean and coastal adaptation projects in the Asia–Pacific region and identify the current funding gap. It is widely acknowledged that tracking efforts focusing on climate change adaptation is challenging because of the absence of globally agreed accounting assumptions (Donner et al. 2016). Therefore, there are few sectoral analyzes of the adaptation financial flow. As for ocean and climate actions, there have been several attempts to track ocean-related projects financed by global climate funds, official development assistance, and grants from philanthropists, but these analyzes focus only on specific sectors (e.g., fisheries) or countries (Guggisberg 2019; Piñeiro-Antelo et al. 2019; Blasiak and Wabnitz 2018; Berger et al. 2019). The author also acknowledges the Sustainable Ocean Economy Database, the tracker for bilateral ocean finance flows developed by the OECD, which provides datasets of ocean-related funding from OECD and partner organizations. This study is the first critical step to track multilateral efforts toward ocean and coastal adaptation. The specific research questions include: (1) what is the increase in the overall size of the fund over the past decade? (2) What is the scale of aid for SIDS? (3) What kind of activities receive more funds by sector and different types of approaches?

The study measures data from four multilateral development banks and funds: the World Bank Group (WBG), which includes the International Bank for Reconstruction and Development (IBRD) and International Development Association (IDA); Global Environment Facility (GEF); Green Climate Fund (GCF), and Asian Development Bank (ADB) over the period of 2010–2019. The targeted multilateral donors were selected based on two criteria, either (1) they financed climate adaptation in the Asia–Pacific countries or (2) climate change adaptation in low-and low-middle-income states is one of their priority areas. WBG and ADB are the first and second largest donors of climate adaptation finance (UNFCCC, 2020), and GEF and GCF are the two operating entities of the UNFCCC financial mechanism, which are mandated to support vulnerable countries' climate change adaptation. The GEF supports climate adaptation through the Least Developed Countries Fund (LDCF) and the Special Climate Change Fund (SCCF). As both funds are established to help vulnerable countries adapt to the negative impacts of climate change, they are not rigidly distinguished in this analysis. LDCF places more emphasis on helping the poorest countries implement National Adaptation Programs of Action, and projects in Asia–Pacific account for only a small portion of its total efforts (Sovacool et al. 2017).

Table 7.1 Search terms for extracting projects from open database

Main terms	Marine/Ocean/Sea/Seascape/Coastal/Coast
Sub terms	Adaptation/Climate/Ecosystem/Sea-Level Rise/Erosion/Disaster Risk/Management/Protection/EbA/Integrated Zone/MPAs/Protected Areas/Communities/Infrastructure/Fisheries/Aquaculture/Tourism/Government/Health/Coral Reefs/Mangrove/Salt Marshes/Wetlands/Estuaries/Sandy Beaches/Dunes

7.4.2 Project Identification

An online database or a project finder of each financial institution provided the status of aid project approvals between 2010 and 2019. The search terms used to extract projects related to ocean and coastal adaptation measures are listed in Table 7.1. Projects that have been cancelled, are located in landlocked countries, and explicitly target inland areas or freshwater ecosystems (such as lakes) are excluded.

Moreover, the project documents are collected to be further analyzed for keyword and phrase frequency using WordStat9, a text analysis software tool to illustrate patterns. Keywords and categories are adjusted before the frequency analysis is performed to specify how the textual information should be processed and to treat variant forms as group-related words under meaningful categories. Words with little semantic value, such as pronouns and conjunctions, were excluded.

7.4.3 Categorization by Sector and Approach

According to the research purpose, each project is classified by the following sectors: water resources, urban infrastructure, fisheries and aquaculture, coastal tourism, ecosystem conservation, community disaster risk management, and monitoring. The classification is based on project descriptions provided by the respective organizations, which are available on their web pages or downloadable project documents.

In addition, the budget allocated to each project is aggregated according to the IPCC SROCC classification of adaptation measures for the ocean and coastal areas (Bindoff et al. 2019). The projects are divided into six adaptation categories based on their objective: physical, ecological, social, governance, economic, and knowledge adaptation. Examples of adaptation options for each category are presented by the author in Table 7.2.

The physical approach addresses the impact of climate change on physical processes, such as coastal infrastructure damage, disruption of urban systems, and land subsidence. Possible adaptation measures include hard and soft engineering responses to protect coastal assets and save lives. This approach is categorized by building infrastructure that increases the resilience of coastal communities, such as

Table 7.2 Typology of ocean and coastal adaptation projects

Categories[1]	Example of adaptation options
Physical	Hard and soft engineering responses including infrastructure development
Ecological	Ecosystem-based coastal management
Social	Community capacity building, stakeholder participation, health
Governance	Policies (strategies, guidelines, and plans), institutional capacity building, monitoring
Economic	Value chain, productivity improvement, investment
Knowledge	Awareness raising, knowledge-sharing

dikes, water supply systems, and flood protection. Early warning systems designed to avoid damage caused by coastal hazards are also included.

The ecological approach includes responses to ecosystem degradation, biodiversity loss, and impacts on habitats and species. Restoration and protection of coastal ecosystems, such as mangroves, coral reefs, and wetlands, are the basis of this approach. The ecosystem-based measures, such as MPA, NbS, EbA, and Eco-DRR, are all included in this approach. These measures not only benefit marine and coastal ecosystems but also bring further benefits to livelihoods through disaster risk reduction, water security, etc.

The social approach measures a wide range of responses because climate change affects societies, communities, and households. For instance, decreased access to ecosystem services, decline in livelihood opportunities, increasing living costs, food insecurity, and health risks may be triggered by climate change. Certain gender-related impacts are also noted as unevenly distributed impacts of climate change. These issues need to be addressed through improvement of social protection, community participation, and empowerment, which are the key elements to build social, economic, and cultural resilience in coastal communities.

The governance approach highlights the capacity challenges to cope with climate risks. Developing appropriate strategies, guidelines, or plans for climate change adaptation is often challenging for institutions with limited financial and human resources. Monitoring climate change and its impact on coastal communities and industries also requires technical capacities. This approach is focused on the improvement of these necessary capacities. As higher governance quality is likely to attract more adaptation funds (Robinson and Dornan 2017), this approach can create a virtuous circle for improving adaptive capacity through improved governance.

The economic approach indicates countermeasures for increased cost, business disruptions, and losses, as well as the decreased value of assets or products due to incremental climate change impacts. The resilience value chain and productivity improvement are typical examples of this approach. In many Asia–Pacific coastal and island countries, marine resources and ecosystems are closely linked with economic activities and livelihoods, such as fisheries, pearl farming, and coastal tourism. The

[1] Bindoff et al. (2019), p 525.

economic approach is to build resilient economic systems against external shocks of climate-related hazards.

The knowledge approach supports decision-making under uncertainty of climate change impacts at the institutional, household, and individual levels. Raising awareness and knowledge-sharing should be the key elements of climate change adaptation strategies at the community level, which often entail science communication, knowledge system development, and stakeholder involvement.

7.5 Results

7.5.1 Trends in Ocean and Coastal Adaptation Development Finance in the Asia–Pacific

As a result of project identification, 139 projects in total are identified. Table 7.3 lists the breakdown of the identified projects as well as the number of available documents utilized for text analysis in Sect. 5.3.

Figure 7.1 shows the total budget (excluding co-financing, government expenditure funds, and indirect costs) of climate change adaptation projects related to ocean and coastal areas approved by the WBG, GEF, GCF, and ADB between 2010 and 2019. The results show that adaptation funding for the oceans and coastal areas of the Asia–Pacific region has increased almost fivefold over the past decade, from US$ 852 million to approximately US$ 4.3 billion. On the other hand, aid to SIDS has also increased significantly, from US$ 5 million to US$ 5.16 million, but the share of the total amount remains low. Table 7.4 shows the top 10 countries that received ocean and coastal adaptation funding over the past decade. The top three countries—India, Bangladesh, and Vietnam—notably, received large-scale loans for projects such as urban infrastructure construction. The Philippines, at fourth place, is particularly high on the list due to a large-scale rural development project by the IBRD, which included the development of value chain infrastructure to increase the

Table 7.3 Number of extracted projects and documents

Financial institution	Identified projects	Available documents	Document type
Asia Development Bank (ADB)	38	38	Project data sheet
World Bank Group (WBG)	27	24	Approved funding proposal, project paper
Global Environment Facility (GEF)	60	55	Project identification form document for WPI
Green Climate Fund (GCF)	14	14	Approved funding proposal
Total	139	131	

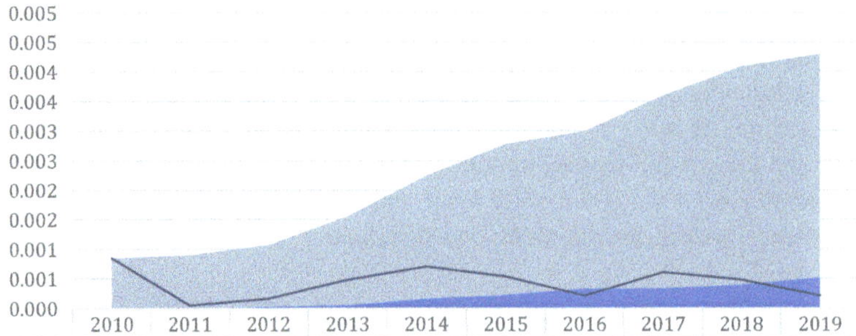

	2010	2011	2012	2013	2014	2015	2016	2017	2018	2019
Total	0.852	0.902	1.067	1.550	2.247	2.778	2.993	3.601	4.084	4.296
SIDS	0.005	0.008	0.022	0.045	0.156	0.221	0.332	0.332	0.393	0.516
Yearly total	0.852	0.050	0.165	0.483	0.697	0.531	0.215	0.608	0.483	0.212

Fig. 7.1 Total approved budget for oceans and coastal adaptation projects in the Asia–Pacific, 2010–2019 (US$ million)

Table 7.4 Top 10 countries which received project aids for oceans and coastal adaptation (including loan and grant) over 2010–2019

	Country	Population (thousands)	Total amount (US$ million)	Amount per capita (US$)
1	India	1,380,004	1433.3	1.04
2	Bangladesh	164,689	825.4	5.01
3	Viet Nam	97,339	616.0	6.33
4	Philippines	109,581	453.3	4.14
5	Timor- Leste	1,318	132.1	100.19
6	Samoa	198	125.8	634.08
7	Multiple countries	–	95.2	–
8	Cambodia	16,719	92.1	5.51
9	Vanuatu	307	92.0	299.65
10	Marshall Islands	59	85.2	1438.56

[a] Population based on the most recent national data available in the World Bank (2021)
Note: "Regional" means a project covers multiple countries

competitiveness of agriculture and fishery products. Meanwhile, countries which ranked sixth and below are far behind. For instance, Samoa, ranked sixth, received a relatively small loan to strengthen its port, while the rest of the funds were mainly grants and technical assistance. Some SIDS, such as Samoa, Timor-Leste, Vanuatu, and the Marshall Islands are ranked high as well, which implies that a small number of countries account for most of the ocean and coastal adaptation finance through loans, with the remainder shared among other countries, including other SIDS.

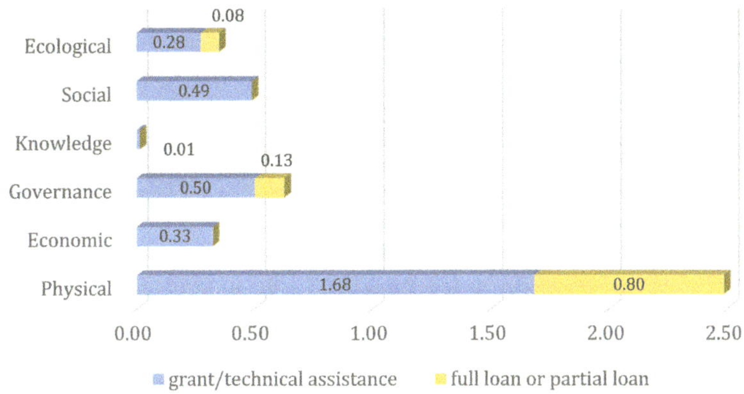

Fig. 7.2 Approved budget for ocean and coastal adaptation projects in the Asia–Pacific by approaches, 2010–2019 (US$ billion)

7.5.2 Sectoral Classification

Figure 7.2 compares the approved budget between each category of adaptation measures. Physical adaptation options show the largest amount (US$ 2.48 billion). Activities targeting governance adaptation received the second largest allocation (US$ 0.63 billion), followed by social adaptation measures (US$ 0.49 billion). Almost identical budget amounts of US$ 0.36 billion and US$ 0.33 billion were allocated to ecological measures and economic measures, respectively. The budget related to knowledge was the lowest at US$ 0.01 billion. The share of loan-based finance is also displayed, which illustrates that all the projects in social, knowledge, and economic categories are financed by grants. One third of the physical adaptation projects, including infrastructure development, are financed by loans. For projects related to ecological and governance adaptation, relatively small but noticeable budget amounts were loan-based.

The number of approved projects for a given period is shown in Fig. 7.3. The greatest share of the ocean and coastal adaptation projects target ecosystem conservation and protection (32%), including EbA, MPA, and Ridge to Reef. Urban infrastructure and fisheries and aquaculture account for 22% and 21% of the total number of approved projects respectively. Projects related to community disaster risk management (DRM) constitute 12%, followed by water resources (10%), and monitoring and evaluation (3%).

7.5.3 Mapping the Project Components

Figure 7.4 represents the association between all words in the collected ocean and coastal adaptation project descriptions between 2010 and 2019. The size of the bubble

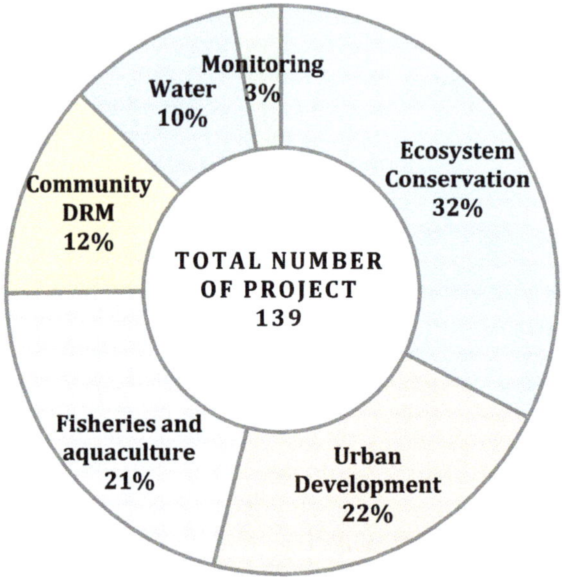

Fig. 7.3 Sectoral share of the approved ocean and coastal adaptation projects in the Asia–Pacific, 2010–2019

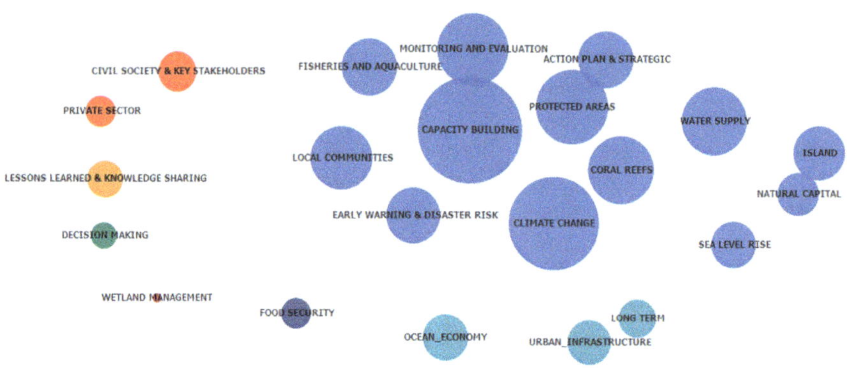

Fig. 7.4 Bubble plot visualizing the trend in ocean and coastal adaptation projects

represents the frequency with which a word in a particular category appears in all project documents over this period, with the most found keywords located closest together within the same paragraph. The distance between words indicates a correlation, and the color represents the cluster. Note that the author adjusted the text analysis processes by creating and modifying dictionaries, as illustrated in the previous section.

The data visualization results illustrate that the most influential factor is capacity building. It also shows that capacity building is mainly associated with monitoring and evaluation, protected areas, action plan development, local communities, early warning and disaster response, fisheries, and coral conservation activities. In the lower part of the figure, there is an observable link between the ocean economy and urban infrastructure, which is often linked to long-term discussions. In addition, knowledge-sharing and food security also appear frequently as well as civil society and key stakeholders, which seem to be accompanied by the involvement of the private sector.

7.6 Discussion

From the results of this study, a cumulative total of adaptation funding for ocean and coastal areas has significantly grown over the past decade, although finance for SIDS accounts for a small share of the total. Previous studies provide the possible reasons for the scarcity of funds for SIDS, such as limited credit-worthiness, inadequate implementation and management capacities at national levels, and unfamiliarity with international adaptation financing architecture (Robinson and Dornan 2017; Quak 2019). The results show, however, that some Pacific SIDS have received noticeable sums of aid per capita: US$ 1,439 for Marshall Islands, US$ 634 for Samoa, and US$ 300 for Vanuatu.

Although the number of projects related to ecosystem conservation is the largest, the scale of investment in ecological adaptation measures is relatively small. Instead, more money is spent on governance-related adaptations, such as government planning or institutional capacity building. This implies that these funds are not used directly but indirectly contribute to ecological adaptation through planning and other policy processes. Hence, it should be noted that the fact that there are many projects bearing the title of ecosystem conservation does not necessarily mean that ecosystem is restored in practice.

With respect to monitoring efforts, the number of projects and the budget scale seems to be limited, even though monitoring and evaluation of capacity development are common components of ocean-related projects implemented in Asia–Pacific. This suggests that monitoring and evaluation are still challenges from a financial resource perspective. Likewise, in terms of approach, further promotion of knowledge-based adaptation measures, including data collection and the establishment of information sharing systems, could be financed at a larger scale. The scale of funding for economic adaptation measures is also relatively small, and it is necessary to create an environment in which communities and small-scale business sectors affected by climate change can access funding and insurance. Very few projects have focused directly on coastal tourism and it is important to further explore the development of climate-resilient tourism. To enhance resilience in vulnerable coastal communities, future blue economy investment flows should be directed to the areas which do not receive sufficient development aid.

Further consideration should made on how blue finance can be integrated into climate change adaptation to fulfill the identified gaps. The perspective of climate change adaptation in the implementation of blue finance must be brought into the mainstream discourse to address the challenges to adaptation in the ocean and coastal zones. There are two possible approaches for achieving this. First, to tackle these specific challenges, a better understanding of ocean and coastal adaptation options and their effectiveness must be developed. As calculating the demand for adaptation finance is generally considered difficult, the adaptation needs in the oceans and coastal areas are also not clear. Therefore, it is important to enhance opportunities for local stakeholders in recipient countries to analyze their own local adaptation needs and access funds through appropriate project formulation. Currently, the need for and benefits of adaptation measures in the ocean and coastal areas are not fully recognized, or the needs are recognized but not translated into projects. Given the recent focus on marine conservation, the awareness of the gap between available resources and project development will become more evident. To manage this situation, it is crucial to provide a toolkit to explain regional climate change impacts, as well as clearly present adaptation options and benefits in the ocean and coastal zones.

Second, it is important to integrate climate change adaptation into the blue economy discourse. In addition to the widespread recognition of climate change as an economic threat, evidence that investment in adaptation measures is essential for securing economic benefits, and a model for "bankable" project formation will be essential. The tracking analysis shows that, with the exception of infrastructure construction, loan investment is overwhelmingly low and that ocean and coastal adaptation projects are basically grant-based. Furthermore, economic adaptation options are limited and except for fisheries and aquaculture, blue economy promotion and climate change adaptation capacity building have not been considered in an integrated manner. Projects such as building value chains and setting up funds would be useful for future reference.

7.7 Concluding Remarks

The status of aid for marine and coastal climate change adaptation in the Asia–Pacific region by using project data from major multilateral banks and global climate funds is investigated. The study finds that there has been a steady increase in the amount of adaptation funds directed to the oceans and coastal areas in the Asia–Pacific region, estimated to grow from US$ 852 million to around US$ 4.3 billion from 2010 to 2019. The allocation of funds focuses on ecosystem conservation, urban development and the fishery sector. Furthermore, physical approaches to climate change based on infrastructure development have attracted international aid the most, while knowledge-based adaptation remains a challenge. The study also reveals that most of the remaining aid has been used for activities related to governance, which suggests that this is the focus of the Asia–Pacific countries with regard to ocean and coastal adaptation. Financing monitoring processes should also be promoted, as they account for only a small portion of total aid for oceans and coastal adaptation activities in the region. Given the growing interest in ocean conservation and sustainability, this study provides guidance on emerging blue finance developments to enhance adaptation capacity and resilience in the oceans and coastal zones. Future investments should cover the shortcomings of existing aid flows and must directly reach the vulnerable. While increasing blue finance initiatives, the international community needs to have a clear vision based on the overall picture that contributes to ocean health and resilient coasts, rather than blindly investing funds.

Acknowledgements This study was partly supported by the Collaborative Research Project of the International Research Institute of Disaster Science, Tohoku University.

References

Agardy T, Davis J, Sherwood K et al (2017) Taking steps toward marine and coastal ecosystem-based management: an introductory guide. United Nations Environmental Program, Nairobi

Azad MAK, Uddin MS, Zaman S et al (2019) Community-based disaster management and its salient features: a policy approach to people-centred risk reduction in Bangladesh. APJORD 29(2):135–160

Becker A, Ng AK, McEvoy D et al (2018) Implications of climate change for shipping: ports and supply chains. Wiley Interdiscip. Rev. Clim. Change 9(2):508

Behlert B, Diekjobst R, Felgentreff C et al (2021) World Risk Report 2020 Focus: forced displacement and migration. Bündnis Entwicklung Hilft and Ruhr University Bochum—Institute for International Law of Peace and Armed Conflict (IFHV)

Bell JD, Cisneros-Montemayor A, Hanich Q (2018) Adaptations to maintain the contributions of small-scale fisheries to food security in the Pacific Islands. Mar Policy 88:303–314

Berger MF., Caruso V, Peterson E (2019) An updated orientation to marine conservation funding flows. Mar Policy 107. http://www.ncbi.nlm.nih.gov/pubmed/103497

Bindoff NL, Cheung WWL, Kairo JG et al (2019) Changing ocean, marine ecosystems, and dependent communities. In: IPCC special report on the ocean and cryosphere in a changing climate

Blanchard JL, Jennings S, Holmes R et al (2012) Potential consequences of climate change for primary production and fish production in large marine ecosystems. Philos Trans R Soc Lond, B, Biol Sci 367(1605):2979–2989

Blasiak R, Wabnitz CCC (2018) Aligning fisheries aid with international development targets and goals. Mar Policy 88:86–92

Cohen-Shacham E, Walters G, Janzen C et al (2016) Nature-based solutions to address global societal challenges. International Union for Conservation of Nature, Gland.

Cooley SR, Bello B, Bodansky D et al (2019) Overlooked ocean strategies to address climate change. Glob Environ Change 59:101968

Dearden P and Manopawitr P (2011) Climate change–coral reefs and dive tourism in South-east Asia. Disappearing destinations: climate change and future challenges for coastal tourism. CAB International, UK, pp 144–160

Donner SD, Kandlikar M, Webber S (2016) Measuring and tracking the flow of climate change adaptation aid to the developing world. Env Res Lett 11(5):054006

Food and Agriculture Organization (2018) The state of world fisheries and aquaculture 2018—Meeting the sustainable development goals. Food and Agriculture Organization of the United Nations, Rome

Gaines S, Cabral R, Free C et al (2019) The expected impacts of climate change on the ocean economy. World Resources Institute, Washington DC

Gattuso JP, Magnan AK, Bopp L et al (2018) Ocean solutions to address climate change and its effects on marine ecosystems. Front Mar Sci 5:337

Giffin AL, Brown CJ, Nalau J (2020) Marine and coastal ecosystem-based adaptation in Asia and Oceania: review of approaches and integration with marine spatial planning. Paci Conse Biol. 27(2):104–117

Guggisberg S (2019) Funding coastal and marine fisheries projects under the climate change regime. Mar Policy 107:103352

Heenan A, Pomeroy R, Bell J (2015) A climate-informed, ecosystem approach to fisheries management. Mar Policy 57:182–192

Hoegh-Guldberg O, Northrop E, Lubchenco J (2019a) The ocean is key to achieving climate and societal goals. Science 365(6460):1372–1374

Hoegh-Guldberg O, Lovelock C, Caldeira K et al (2019b). The ocean as a solution to climate change: five opportunities for action. High-level Panel for a Sustainable Ocean Economy https://oceanpanel.org/sites/default/files/2019b-10/HLP_Report_Ocean_Solution_Climate_Change_final.pdf.

Inaotombi S, Mahanta PC (2018) Pathways of socio-ecological resilience to climate change for fisheries through indigenous knowledge. Hum Ecol Risk Assess 25(3):1–13

IPCC (2019) Summary for policymakers. In: Pörtner H-O, Roberts DC, Masson-Delmotte V et al (eds) IPCC special report on the ocean and cryosphere in a changing climate (in press)

Johnson JE, Bell JD, Gupta AS (2015) Pacific islands ocean acidification vulnerability assessment. Pacific Regional Environment Programme, Apia.

Khan A, Amelie V (2015) Assessing climate change readiness in Seychelles: implications for ecosystem-based adaptation mainstreaming and marine spatial planning. Reg Env Change 15(4):721–733

Kittinger JN (2013) Human dimensions of small-scale and traditional fisheries in the Asia–Pacific region. Pac Sci 67(3):315–325

Lam JSL, Liu C, Gou X (2017) Cyclone risk mapping for critical coastal infrastructure: cases of East Asian seaports. Ocean Coast Manag 141:43–54

Lam VW, Allison EH, Bell JD et al (2020) Climate change, tropical fisheries, and prospects for sustainable development. Nat Rev Earth Environ 1(9):440–454

Lauria V, Das I, Hazra S et al (2018) Importance of fisheries for food security across three climate change vulnerable deltas. Sci Total Environ 640:1566–1577

MDBs (2019) 2018 Joint report on multilateral development banks' climate finance. Multilateral Development Banks Group

MDBs (2021) 2020 Joint report on multilateral development banks' climate finance. Multilateral Development Banks Group

Nara P, Mao G, Yen T (2014) Applying environmental management policy for sustainable development of coastal tourism in Thailand. J Environ Prot Ecol 2(1):19–23

OECD (2016) The Ocean Economy in 2030, OECD Publishing, Paris, https://doi.org/10.1787/9789264251724-en

OECD (2021) Climate finance provided and mobilised by developed countries: aggregate trends updated with 2019 data, climate finance and the USD 100 Billion Goal. OECD Publishing, Paris. https://doi.org/10.1787/03590fb7-en

Oliver P, Clark A, Meattle C (2018) Global climate finance: an updated view 2018. Climate policy initiative. Available at https://www.climatepolicyinitiative.org/wp-content/uploads/2018/11/Global-Climate-Finance-_-An-Updated-View-2018.pdf. Accessed 18 Dec 2021

O'Mahony C, Gray S, Gault J et al (2020) ICZM as a framework for climate change adaptation action experience from Cork Harbour. Ireland Mar Policy 111:102223

Pandey BH, Okazaki K (2005) Community-based disaster management: empowering communities to cope with disaster risks. Reg Dev Dialogue 26(2):52

Pauw WP (2015) Not a panacea: private-sector engagement in adaptation and adaptation finance in developing countries. Clim. Policy 15(5):583–603

Piñeiro-Antelo M, Villa FRD, Santos XM (2019) ODA in Galicia (Spain). The importance of the fisheries sector and the cultural priority. Mar Policy, 107, http://www.ncbi.nlm.nih.gov/pubmed/103455.

Quak E (2019) How losing access to concessional finance affects small island developing states (SIDS) K4D Research Helpdesk Report. Institute of Development Studies, Brighton

Rahman S, Rahman MA (2015) Climate extremes and challenges to infrastructure development in coastal cities in Bangladesh. Weather. Clim. Extremes 7:96–108

Robinson S, Dornan M (2017) International financing for climate change adaptation in small island developing states. Reg Environ Change 17(4):1103–1115

Saghir J, Schaeffer M, Chen A et al (2021) State and trends in adaptation report 2020. Global Center on Adaptation, Groningen

Smith-Godfrey S (2016) Defining the blue economy. MaritE Aff: J Natl MaritE Found India 12(1):58–64

Sovacool BK, Linnér BO, Klein RJ (2017) Climate change adaptation and the Least Developed Countries Fund (LDCF): qualitative insights from policy implementation in the Asia–Pacific. Clim Change 140(2):209–226

Spalding MD, Ruffo S, Lacambra C et al (2014) The role of ecosystems in coastal protection: adapting to climate change and coastal hazards. Ocean Coast Manag 90:50–57

Sumaila UR, Walsh M, Hoareau K et al (2021) Financing a sustainable ocean economy. Nat Comm 12(1):1–11

Sumaila UR, Walsh M, Hoareau K et al (2020) Ocean finance: financing the transition to a sustainable ocean economy. World Resources Institute, Washington DC

UN Environment Programme (2021a) Adaptation gap report 2020. United Nations Environment Programme, Nairobi. https://www.unep.org/resources/adaptation-gap-report-2020. Accessed 18 Dec 2021a

UN Environment Programme (2021b) Turning the tide: how to finance a sustainable ocean recovery. United Nations Environment, Nairobi

UNFCCC (2016) Biennial assessment and overview of climate finance flows report, United Nations Framework Convention on Climate Change Standing Committee on Finance, Bonn. https://unfccc.int/sites/default/files/resource/54307_1%20-%20UNFCCC%20BA%202020%20-%20Report%20-%20V4.pdf

De Vos K, Hart B (2020) The ocean finance handbook. Friends of Ocean Action

Ward RD, Friess DA, Day RH et al (2016) Impacts of climate change on mangrove ecosystems: a region by region overview. Ecosyst Health Sust 2(4):e01211

Wong E, Jiang M, Klint L et al (2013) Policy environment for the tourism sector's adaptation to climate change in the South Pacific—the case of Samoa. Asia Pac. J. Tour. Res. 18(1–2):52–71

World Bank and United Nations Department of Economic and Social Affairs (UNDESA) (2017) The potential of the blue economy: Increasing long-term benefits of the sustainable use of marine resources for small island developing states and coastal least developed countries. World Bank, Washington DC

World Bank (2021) World Development Indicators database, *Population 2020*. https://databank.worldbank.org/data/download/POP.pdf. Accessed 18 Dec 2021

Chapter 8
Early Recovery in Humanitarian Response—In the Cases of Four Disasters in the Asia-Pacific Region

Tsukasa Katsube

Abstract The Sendai Framework for Action 2015–2030 and its Priority 4 state that effective response work should be well-integrated into disaster risk reduction concepts, and that preparing ahead is crucial toward achieving "Build Back Better." The same concept was highlighted in the United Nations General Assembly Resolution 46/182, and endorsed in 1991 as the "Continuum from relief to rehabilitation and development." For the past 30 years, the concept was at the heart of humanitarian response work, and response systems and methodologies were developed accordingly. This chapter focuses on the early phase of disaster response and how the abovementioned concepts have been incorporated from the perspective of the early recovery element in humanitarian financing, and the early recovery cluster in the humanitarian cluster system as a key driver that ensures the effective transition from response to rehabilitation and recovery. Four recent natural disasters in the Philippines, Vanuatu, Nepal, and Indonesia will be compared to explore how early recovery was positioned in the early response phase. Focusing on early recovery while studying humanitarian financing will help us understand how humanitarian response work remains relevant for Asian countries and enables them to steadily adapt to the growing risk of natural disasters caused by climate change.

Keywords Early recovery · Emergency response · Humanitarian financing

8.1 Introduction

When a natural hazard triggers a disaster in a country, the responsibility to take care of the population rests primarily with the government of that country. In the aftermath of such an occurrence, it is largely a humanitarian response that saves lives and alleviates the suffering of the affected people. Likewise, the recovery and reconstruction of society and the livelihoods of the affected also take place in parallel with response activities from the day of the inception of the disaster event. The importance of

T. Katsube (✉)
Japan International Cooperation Agency, Tokyo, Japan
e-mail: katsube.tsukasa.3@jica.go.jp; tsukasakatsube@yahoo.com

M. Ishiwatari and D. Sasaki (eds.), *Financing Investment in Disaster Risk Reduction and Climate Change Adaptation*, Disaster Risk Reduction,
https://doi.org/10.1007/978-981-19-2924-3_8

incorporating longer-term aspects even during the earliest phases of humanitarian response has been stressed upon for decades, and terms such as the humanitarian-development nexus, continuum, and seamless approach have been discussed in many reports. However, it still begs the question of how this is realized in the practice of humanitarian response. This article focuses on humanitarian aid primarily from the perspective of funding, and highlights the early recovery of the society that is possible due to humanitarian response.

A humanitarian response is ideal if the required funding is secured by the host government. However, such expenses are often too large for one country to cover and reliance on the international community is inevitable. The key questions are how much money is required for the response activities, who outlines the whole amount and decides how much, and how this funding is financially sourced. The answers to these questions will be revealed through the processes of assessment and analysis (A&A), reporting, and financial appeal. Hence, the first part of this chapter will focus on the standardized international methodologies required for a series of financial mobilization processes for humanitarian response.

Later, we will focus on the early recovery of the humanitarian cluster system to understand how humanitarian funding has been spent on bridging humanitarian and development work. This trend will be identified by comparing four natural disaster cases—Typhoon Haiyan (2013) in the Philippines, Cyclone Pam (2015) in Vanuatu, the Gorkha Earthquake (2015) in Nepal, and the Sulawesi Earthquake (2018) in Indonesia.

8.2 Humanitarian Leadership, Humanitarian Country Team, and the Humanitarian Cluster System

Before the assessment and analysis discussion, it is necessary to understand the basics of humanitarian response architecture. Three elements are essential, especially in the context of assessment. These include humanitarian leadership, the humanitarian country team (HCT), and the humanitarian cluster system.

8.2.1 Humanitarian Leadership, Resident Coordinator, and Humanitarian Coordinator

The resident coordinator (RC) is the highest senior representative of both the UN and the non-UN humanitarian and development-related actors. The role of the RC was highlighted by the United Nations General Assembly (UNGA) Resolution 46/182[1] (GAR46/182) in 1991 as a key coordination function of a country-level humanitarian response. The role was further empowered in the 2018 UNGA Resolution

[1] United Nations General Assembly Resolution 46/182, 19 Dec 1991.

72/279[2] for its development-focused coordination function. Likewise, the Humanitarian Coordinator (HC) is responsible for the country-level coordination of humanitarian response. Both the HC and RC can be the same person. In this case, the RC either takes up the role of the HC when a humanitarian situation arises, or he assigns other senior personnel who are more capable in dealing with humanitarian response. Thus, the RC or the HC is a dedicated senior position that coordinates both humanitarian responses as well as the development effort following a disaster.

8.2.2 Humanitarian Country Team (HCT)

The Humanitarian Country Team (HCT) is composed of both UN and non-UN in-country humanitarian responders and national and international actors. IASC defines that "The HCT's overall goal is to ensure that inter-agency humanitarian action alleviates human suffering and protects the lives, the livelihoods, and dignity of people in need[3]", and "as the top inter-agency humanitarian leadership body in a country, the HCT's primary purpose is to provide strategic direction for collective inter-agency humanitarian response.[4]" It also "makes decisions to ensure that country-level humanitarian action is well-coordinated, principled, timely, effective, and efficient.[5]" The establishment of the HCT is based on the decision of RC or HC in consultation with relevant actors, including the Humanitarian Response Coordinator, the head of the UN Office for the Coordination of Humanitarian Affairs (OCHA), and the OCHA itself. The HCT is normally established when a humanitarian crisis erupts and an HC is appointed, or when a protracted humanitarian crisis exists.

8.2.3 The Humanitarian Cluster System

The humanitarian cluster is a coordination system that ensures that the response effort is conducted in a coordinated manner. It was originally introduced in the Humanitarian Reform of 2005, and further sophisticated in the Transformative Agenda of 2011, and so on. The system is activated when "response and coordination gaps exist due to a sharp deterioration or a significant change in the humanitarian situation."[6] The global cluster system has 11 pre-determined needs areas with cluster lead agencies. When disaster strikes in a country and the situation leaves gaps in humanitarian needs, the humanitarian clusters will be activated only within clusters that are in need. Thus, all 11 clusters may not be established.

---

[2] United Nations General Assembly Resolution 72/279, May 2018.

[3] IASC (2017), "Standard Terms of Reference for Humanitarian Country Teams".

[4] Ibid.

[5] Ibid.

[6] IASC (2015), "Reference Module for Cluster Coordination at Country Level".

8.2.4 *UNDAC*

The United Nations Disaster Assessment and Coordination (UNDAC) is OCHA's coordination tool. It sends coordination experts to the affected country following government consent. The UNDAC consists of a pool of global experts, and most members normally work for their respective governments. The UNDAC team is on call when a major disaster strikes and the affected government, RC or HC requests the booster capacity for humanitarian response coordination. The UNDAC serves the affected government as well as the RC/HC to support the field coordination led by the affected government. UNDAC members have profound knowledge and experience in field coordination and generally work with the government, RC/HC, and HCT, and facilitate the early stage of assessment, reporting, launching appeals, and so on in order to orchestrate coordination activities.

8.3 Initial Humanitarian Response Planning

8.3.1 *Flash Appeal (FA)*

In the early stage of disaster response, suitable planning must take place to deliver the humanitarian response efficiently and effectively. The primary planning decision maker is the host country, but they are likely to cooperate with the RC/HC and HCT. In the standard humanitarian response methodology set by the Inter-Agency Standing Committee (IASC), the plan is called the Humanitarian Response Plan (HRP). The HRP should be the result of a coordinated assessment and should be a collaboration with the host government, RC/HC, HCT, and relevant stakeholders.

The flash appeal (FA) is part of the HRP and takes place in the earliest stage of the response when information about the disaster impact is limited. It is most likely drafted based on assessment information conducted by the host government and other actors. The nature of the FA is described as "quick and dirty," meaning that it maximizes speed by compromising on a certain level of accuracy, but it is good enough to outline the initial surge of humanitarian planning for the first 6 months. The FA consists of the country strategy, strategic objectives, response strategy, and sector strategies[7] and is expected to be released within the first seven days. The primary information it provides includes the needs for each humanitarian cluster area as well as the rationale behind the needs, and funding requirements. The funding requirements include those for individual projects and the total sum of the requirements.

[7] OCHA, "UNDAC Field Handbook", Initial response planning and Flash Appeal, Chapter L, p. 25.

8.3.2 Humanitarian Response Plan (HRP)

Even after the FA is released, it is assumed that assessment efforts continue to identify potential humanitarian needs that were missed out from the first assessment, and reach out to deeper targets, such as geographical areas that had not yet been accessed. Additionally, the response environment and humanitarian needs are likely to change dynamically over time. Considering these expected changes in other variables, such as the deterioration of safety and security situations, the humanitarian response plan must stay updated. Thus, the FA needs to be revised accordingly. The actual revision comes within four to six weeks since the occurrence of the disaster, and the updated version is no longer called the Flash Appeal but the Humanitarian Response Plan (HRP). This gauges the humanitarian requirement for the next one to three years as the result of the adoption of a more thorough methodology, called multi-sector initial rapid assessment or MIRA.

8.4 Assessment and Analysis Process

The initial phase of the assessment is mostly done by desktop-based secondary data collection by various actors in the affected country, including the host government. In the first few days, the field-type assessment may be conducted, but it mostly takes place spontaneously and is not coordinated. Since the FA is produced as a result of the initial assessment; it may not be based on a thorough knowledge of the disaster situation. The MIRA, however, is expected to be an in-depth assessment with a combined approach, as opposed to a spontaneous one.

The MIRA consists of two major processes: secondary data collection done using a desktop-based search and primary data collection done by field assessment. The design process takes two weeks, which includes three days for secondary data collection, seven days for primary data collection, and four days for joint analysis and reporting.

While the secondary data collection is ongoing, the primary data collection is prepared by various actors in the country, including the host government, the HCT, and other humanitarian actors. This process is ideally commenced by the host government and is generally planned under the leadership of the government and supported by the RC/HC. Once all the actors agree to conduct joint assessments, joint planning for primary data collection takes place. By then, data gaps should be identified based on the results of secondary data collection. Such gaps can be seen in particular geographical areas, subject areas, or groups in vulnerable situations.

It should be noted here that the MIRA is not designed to aim for microscopic data. Rather, UNDAC Field Handbook mentioned that it is meant "to develop an initial common understanding of the most pressing priority needs, affected areas, and main affected groups"[8] and this situational overview "aid(s) decision-makers in identifying priority areas during the first phase of the response."[9]

During the preparation for primary data collection, a common methodological framework should be agreed upon among all stakeholders involved in the assessment. This includes where to go, what to know, who to meet, how to do, when to do, and so on. Furthermore, the assessment teams should be formed with multi-cluster/inter-disciplinary groups from the host government (line ministries of particular sectors), the UN, NGOs, the Red Cross, the affected population, and the private sector. The primary data collection is a blend of key informant interviews, direct observation, and community group discussions.

The collected data were processed using analysis, that is, through describing, explaining, interpreting, and forecasting. During the description, all the collected data were summarized and validated, resulting in the revelation of generalized trends. The next is an explanation, in which the "whys" of the identified trends, including cause and effect, relationships and correlations of dates are discussed. Interpretation examines the meaning of the identified data by "attaching meaning to data and evaluating the evidence." Lastly, forecasting allows the anticipation of "what happens next" to develop likely consequences that are expected based on the analysis.[10] The data collected by the primary data collection method transforms into information following evaluation and interpretation and is then consolidated into a MIRA report.

8.5 Reporting and Appeal

In the previous section, the FA, HRP and MIRA methodologies were introduced as likely response procedures that follow a disaster. In recent practice, however, HRP has not been done in response to major natural disasters, but is more often produced in complex emergencies. Instead, during disasters such as Typhoon Haiyan in the Philippines (2013), Cyclone Pam in Vanuatu (2015), Nepal Earthquake (2015), and Sulawesi Earthquake (2018), FAs were produced to cover financing for the first six months of life-saving effort, and was mostly followed up by a Post-Disaster Needs Assessment for the next couple of years for rehabilitation and reconstruction purposes. Hence, we may exclude HRP from the current analysis and focus only on FA.

---

[8] ibid., Chapter I, p11.

[9] Ibid.

[10] IASC (2015), Multi-Sector Initial Rapid Assessment Guidance, P18.

8.5.1 Flash Appeal as a Financial Mobilization Tool

As its name suggests, the primary utilization of the FA is its financial appeal to donors that helps to mobilize the necessary budget. When the initial assessment is concluded and FA is produced, a formal launching session hosted by the government and supported by HC/RC takes place. This is the moment when the world is made aware of the budget that is required for response assistance based on the formal announcement made by the affected government. During the session, embassies and official missions of other countries are usually invited and asked for financial contributions. In response to this appeal, donor countries pledge their contributions within a few days or months.

The FA for the Nepal Earthquake response in 2015, for example, was 18 pages long and consisted of complete information about what is required, why it is required, and how much is required. On the top of the report, it discussed the immediate requirement of almost USD 415 million from the total funding for the purpose of life-saving and protection activities over the course of three months. This was followed by a breakdown of the request based on each humanitarian cluster, where the amount of required funding was specified for a particular cluster (e.g., USD 50,000,00 for Shelter and Non-food Items (NFIs)). The details of the needs and required response actions in each cluster are further explained later, often supplemented by a project list and project details. The listing of project details appears to be a unique component of the FA. It informs readers of the name of the appealing agency, project title, project code, objectives, location, number of people targeted, and required budget.

The FA provides not only the initial common situation awareness, but also the details of humanitarian response needs with financial information. This information may help donors envisage how their financial support can contribute to the response effort in line with the coordinated response direction. Furthermore, the FA serves as a tool for donors to ensure accountability, and justifies their investment behavior by discussing the rationality behind why it is necessary to provide financial support for particular clusters and projects.

The last page of the FA is a reminder that the FA is a financial mobilization tool. It has information on the "Guide to Giving," where it refers to methods of financial contribution by means of three options: 1. Contribution to Flash Appeal, 2. In-kind Relief Aid, 3. Donations through the Central Emergency Response Fund (CERF).

8.6 Financing Sources

8.6.1 Central Emergency Response Fund (CERF)

Out of the multiple methods of financial contribution during the time of natural disasters, the Central Emergency Response Fund (CERF) is one of the oldest and most well-known. It was first established as the Central Emergency Revolving Fund

(CERF) in 1992 as a result of GAR 46/182, and was revised as the currently known CERF in 2005. The CERF grant is reimbursed at the very first stage of humanitarian response, based on a condition called "life-saving criteria".[11] This is an emergency pooled fund that kickstarts the swift execution of humanitarian response activities. While the FA provides a list of items that need to be financially covered, the CERF is one of the financial sources that fill some of the financial requirements.[12]

The application process of the CERF grant starts at the early stages of response in parallel with the FA development process. In general, the HC/RC has the overall authority to manage the CERF for a particular response. The application process is in collaboration with the HC/RC and the HCT, together with each cluster lead in the country. UN organizations are eligible to be direct recipients, whereas NGOs are not. However, NGOs are indirect recipients as implementing partners of UN organizations.[13]

8.6.2 Humanitarian and Bilateral Development Donors

Funding from governments such as those in OECD countries is one of the major financial sources for humanitarian response. A report indicated that the largest funding for humanitarian response came from OECD and DAC countries (USD 21.6 billion, 70%), followed by others (USD 2.5 billion, 8%), and private sectors (USD 6.8 billion, 22%) in 2019.[14] In general, this funding includes all complex emergencies and refugee-involving situations. Particularly for natural disasters, another report studied nine natural disaster-related crises for the period of 18 months from their occurrence, and indicated that, on a grant commitment basis, humanitarian and bilateral donors provided 64% (USD 1,984 million) of the amount with 1,610 projects out of the total committed funding.[15] This shows that bilateral donors still hold the topmost position among humanitarian actors during major natural disasters that trigger international resource mobilization, whereas others such as the business sector, charity-based organizations, affected local governments, and so on, play critical roles in some contexts.

---

[11] CERF Secretariat (2011), Application of the CERF Life-Saving Criteria -An Update.

[12] op. cit., ARES 46/182, and United Nations General Assembly Resolution 60/124, Dec 2005.

[13] OCHA (2018a, b, c), CERF Handbook.

[14] Development Initiatives (2021), Global Humanitarian Report 2021, Humanitarian Assistance in Numbers, pp. 10–11,

[15] Elle C et al. (2021). Funding Disasters: Tracking Global Humanitarian and Development Funding for Response to Natural Hazards, Sect. 4 Size and Types of Flow, p19, Table 3, Funding by type of institution. "1610 projects" was originally appeared as "1468 by Humanitarian donors" and "142 by Bilateral development donors", and Katsube added both together.

8.7 Focus on Early Recovery as the Bridge Between Humanitarianism and Development

A study showed that 69.9% of the "Disaster-Related International Financing" flows into emergency response, with 24.8% going into reconstruction and 3.6% used for prevention and preparedness.[16] The Emergency Response "saves today" using life-saving emergency services, which is undoubtedly of utmost priority for the affected. However, equally important is the need to support the affected for them to be able to envision their "tomorrow", that is, the long road toward the reconstruction of their lives. This perspective should be embedded into all humanitarian works, and this is the heart of the nexus between humanitarian response and development work. In the current humanitarian cluster system, it appears that this approach has been well reflected as an early recovery concept.

Among the 11 preset humanitarian clusters, early recovery led by the UNDP is used as the cluster lead agency. This cluster provides the means to connect emergency humanitarian responses and long-term rehabilitation, recovery, and reconstruction work. According to a Global Cluster for the Early Recovery report, "Early recovery is both an approach and a set of specific actions to help people move away from dependence on humanitarian relief toward development", and early recovery builds resilience during the humanitarian response.[17] The importance of the "humanitarian-development-peace nexus" has been well reiterated countless times and one of the oldest was seen in the GAR46/182 in 1991. Furthermore, the Sendai Framework for Disaster Risk Reduction in 2015 mentioned that well-integrated disaster risk management that includes response is of great importance.[18]

In the following section, case studies of early recovery will be explored with reference to relatively recent disasters that have occurred in the Asia-Pacific region, especially from the perspective of how the funding was sourced for each kind of assistance project.

8.8 Case Studies

Four case studies were selected from recent disaster events, all of which occurred in the Asia-Pacific region. These were Typhoon Haiyan (2013) in the Philippines, Cyclone Pam (2015) in Vanuatu, Gorkha Earthquake (2015) in Nepal, and Sulawesi Earthquake (2018) in Indonesia.

There are several reasons for selecting these four cases. First, these disasters occurred in recent years following the Transformative Agenda, one of the key milestones of humanitarian reform processes wherein some humanitarian response

[16] The Government of Japan, GDFRR, The World Bank GFDRR (2012), The Sendai Report, Managing Disaster Risks for a Resilient Future, p30.

[17] Global Cluster for Early Recovery (2015a, b), Implementing Early Recovery, p10.

[18] Sendai Framework for Disaster Risk Reduction 2015–2030 (2015).

Table 8.1 Details of flash appeals in the four cases

	Haiyan	Haiyan	Vanuatu	Nepal	Sulawesi
YR of occurrence	2013	2013	2015	2015	2018
Hazard type	TC	TC	TC	EQ	EQ
Appeal type	FA	SRP	FA	FA	ERP
Appeal issued (after)	4 days	49 days	11 days	4 days	6 days
Total required (USD)	348 M	775.7 M	37.7 M	422 M	50.5 M
Total sourced	–	60.6%%	58.0%	66.7%	60.8%
ER requirement (USD)	20 M	115.132 M	2.3 M	16 M	2 M
ER proportion/total	5.7%	14.8%	6.0%	3.8%	3.9%
ER requirement met	–	–	0.0%	56.5%	39.0%
Oder in all clusters	7	3	9	7	8
CERF for ER	No	Yes	No	No	Yes

*ER: Early recovery

systems were reviewed, reevaluated, and reinvented.[19] Second, all the disasters occurred in Asia, apart from the case of Vanuatu, which is in the Pacific region. Thus, the study may review commonalities among the cases in Asia and their differences with the disaster in the Pacific. Third, these disasters experienced a similar humanitarian landscape, including the deployment of the UNDAC, establishment of humanitarian clusters, and provision of the FA. These elements are essential for conducting this comparative study. When it comes to the comparison of these cases, the focus will be on the identification of trends in financing during the emergency response phase, especially during early recovery. Verifying each FA is important to ensure conditional equality among the four cases. For details, please refer to Table 8.1.

8.8.1 *Typhoon Haiyan (2013) in the Philippines*

Typhoon Haiyan, locally called Yolanda, made landfall early in the morning of November 8, 2013. Even before landfall, a UNDAC team was deployed to Tacloban, the worst hit area, where it conducted the initial assessment together with the local government and HCT. The assessment resulted in an overview and first evaluation of the magnitude of the damage, and accordingly, the FA was launched on November 12, four days after the occurrence. The total affected population was 11.3 million, and the funding requirement was an estimated USD 301 million. The government initially identified and announced that the following were the most immediate priorities: food, water, sanitation and hygiene, shelter, medicine, debris clearing, and

[19] IASC, IASC Transformative Agenda. https://interagencystandingcommittee.org/iasc-transformative-agenda. Accessed 9Jan 2022.

logistics.[20] Immediately after the occurrence, a total of 14 clusters were established which included security, livelihood, and so on, other than the 11 preset cluster areas.

From the required USD 301 million, the largest funding, approximately one third of the entire requirement, went to Food Security and Agriculture (USD 100 million). This was followed by the emergency shelter, livelihood, education, WASH, and health. Early recovery was ranked seventh among the 14 clusters (USD 20 million). The project proposed under Early Recovery was "Critical Debris Management" by UNDP[21] as part of the FA. The FA was revised 10 days after the original issue, and the total sum was increased to USD 348 million, without any major changes in the composition of the breakdown.

While the FA was launched and kickstarting funding was being sourced, the MIRA was conducted in parallel, and further details were revealed. The FA was revised based on the MIRA output and resulted in the production of the revised FA 49 days after the occurrence, and entitled the Strategic Response Plan (SRP).

The total requirement of the SRP increased to USD 776 million, and this plan was meant to support humanitarian work for a year until November 2014. Remarkably, the requirement for the Early Recovery jumped to approximately five times greater than the original FA, implying that the needs dynamically shifted from urgent physical life-saving requirements of individuals, which was the most pressing factor when the original appeal was issued, to one with a wider interpretation which included not only the individual's livelihood but also the lives of the entire society.

8.8.2 Cyclone Pam (2015) in Vanuatu

Cyclone Pam, a category 5 cyclone struck Vanuatu on March 13, 2015. The hardest hit areas were Shefa, Tafea, Penama, Malampa, and Torba. The cyclone caused five fatalities and affected 166,600 people. The UNDAC was deployed with the consent of the Vanuatu government. It supported the initial assessment led by the government, resulting in the production of the FA 11 days after the disaster. The total requirement was USD 37.7 million,[22] which was the sum of nine Clusters. The primary humanitarian issues identified were in five areas — drinking water, access to food, emergency shelter, medical facilities and supplies, and access to affected people in isolated islands.

According to the Financial Tracking Service (FTS), USD 700,000 was required for the early recovery cluster while USD 2.3 million was identified as the funding

[20] OCHA (2013), Addendum, Philippines: Typhoon Haiyan Action Plan 22 November 2013.

[21] ibid, Project No. PHI-13/ER/61283/776, Time Critical Debris Management in Areas Devastated by Typhoon Haiyan

[22] Flash Appeal, Emergency Response Plan for Vanuatu, Tropical Cyclone Pam, March–June (2015).

required for the FA.[23] Based on the requirements of the FA, the funding was expected to cover "debris clearance, rehabilitation of community rainwater harvesting, and early warning systems for the protection of gardens. Activities included the restoration of livelihoods, especially for young people, women, and vulnerable populations in urban informal settlements and fisheries communities. The interventions also covered the rehabilitation of essential community infrastructures, such as local roads, marketplaces, pedestrian tracks, bridges, and public or cultural buildings (churches and community halls) and community solar energy systems.[24]"

8.8.3 Gorkha Earthquake (2015) in Nepal

A major earthquake with a magnitude of 7.8 struck the Gorkha region in Nepal on April 25, 2015. It claimed 8,659 lives and injured more than 100,000 people. The UNDAC was deployed and worked to support the initial assessment together with stakeholders in the country. The assessment report was turned into the FA, which was launched four days after the occurrence. The report revealed the magnitude of humanitarian response needs and evaluated the required amount of funding necessary. The total amount required was USD 422 million, most of which was required for food security as well as shelter and non-food items (NFI), both of which consist of approximately half of the total requirement. This was followed by WASH, health, logistics, education, early recovery, and so on, making early recovery the seventh largest among the 12 established clusters. For early recovery, USD 16 million was required, and there were three main objectives of use. These were debris management, reconstruction of community infrastructure, and restoration of critical local services. Twenty two projects were already listed as appeals, of which eleven were to be sourced to UN agencies such as UNDP, UNICEF, UN-HABITAT, ILO, UNOPS, and IOM. Other than the three focused areas, there were projects by UNESCO and UNDP that involved the restoration and preservation of relics because the earthquake damaged world heritage sites.[25]

8.8.4 Sulawesi Earthquake (2018) in Indonesia

A series of strong earthquakes hit central Sulawesi Province and were followed by a tsunami on September 28, 2018. Both hazards caused liquefaction as well as landslides and left significant damage in the region, which resulted in 1,581 deaths, over 2,500 people who were seriously injured, and 330,000 people who were left

[23] OCHA, Financial Tracking Service. https://fts.unocha.org/appeals/483/summary. Accessed 9 Jan 2022.

[24] op. cit., Emergency Response Plan for Vanuatu, p11.

[25] Flash Appeal for Response to the Nepal Earthquake, April-July (2015).

homeless. The government and HCT conducted an initial assessment, and the project was titled the Earthquake Response Plan (ERP). This plan appealed for a total USD 50.5 million for the 12 established humanitarian clusters, including early recovery. Early recovery demanded the eighth largest amount of funding (USD 2million), followed by shelter, health, logistics, camp- management, WASH, food security and livelihood, and gender based violence (GBV).

According to the early recovery breakdown, the primary focus during the initial period of three months was on debris clearance and management of heavy equipment, community rehabilitation, emergency employment for manual work, debris disposal sites, rapid recovery of agriculture and non-agricultural livelihoods, and productive assets (e.g., small-scale business, market/trading places, fishing boats, and livestock).[26]

8.9 Analysis and Findings

Some trends regarding early recovery were identified during the analysis of these four cases. First, regardless of the total amount required, the requirement for the early recovery cluster in all the four cases was within the range of 0.04% to 0.06% of the entire requirement. Given that the largest cluster requirement sometimes occupied a quarter of the entire amount, such as the Food Security cluster in the Nepal case, it is obvious that early recovery is likely to be less significant in the humanitarian response stage, especially between three to six months, which the FA intends to cover.

Second, while food security is always a major cluster that requires the largest portion of the FA; being marked the largest in three out of four cases; early recovery is always placed between the seventh to ninth positions in all cases. Compared to other clusters such as education, logistics, camp-management, and camp-coordination, which are likely to be case-dependent with the scale of the requirement not necessarily being predictable, early recovery seems to remain at a comparably mid-lower position within the entire cluster system.

Third, there is no reasonable trend in the pattern of the proportion of the total requirement and sourced/met amount in the early recovery cluster. In the Nepal case, 56.5% of the requirement was sourced,[27] whereas only 39% was met in the Sulawesi case[28] and 0% in the Vanuatu case. However, this does not seem to be relevant to the degree of met/unmet proportion in the total requirement of all clusters in each case.

---

[26] OCHA (2018a, b, c), Central Sulawesi Earthquake Response Plan (Oct 2018–Dec 2018).

[27] OCHA, Financial Tracking Service. https://www.fts.unocha.org/appeals/486/summary. Accessed 9Jan 2022. The FA required 16,049,800 and 56% (sourced 9,070,396) was met.

[28] OCHA, Financial Tracking Service. https://www.fts.unocha.org/appeals/669/summary. Accessed 9Jan 2022. The plan required 2,000,000 and 39% (sourced 785,760) was met.

Fourth, the likely project of early recovery in the humanitarian response phase is debris clearance and management, common in earthquake settings and tropical storm settings in all cases.

Fifth, CERF does not seem to be a likely source of early recovery. As mentioned above, there is a unique item called "Life Saving Criteria" of CERF. In the original discussion, any project under the area of early recovery was neither accepted nor listed under the "activities eligible for CERF funding" based on the definition of "life saving" according to CERF. CERF defines life saving as that which is done "within a short time span, as a remedy, to mitigate or avert direct loss of life and harm to people, and to protect their dignity".[29] However, in the "Application of the CERF Life-saving Criteria- An Update" by CERF Secretariat in 2011, it was mentioned that "early recovery is now a well-accepted component of humanitarian response and, as such, is an integral part of the response from the beginning of a crisis," and that "certain degree of flexibility has been applied."[30] With these histories taken into consideration, the only example of CERF use among the four cases is seen in the Sulawesi response. This was the "Hazard and Enabling the Provision of Critical Emergency Services Through Dignified Work"[31] by the UNDP, and the sourced amount was USD 785,760.

Lastly, early recovery may have received greater attention in the Strategic Response Plan (SRP) for the Haiyan response. The comparison of four cases was made based on their flash appeal. As mentioned above, only the Haiyan case had an SRP update 45 days after the first FA was launched. From the FA to the SRP, the total amount jumped significantly from USD 348 million to USD 775million (223% increase), and the early recovery funding increased accordingly from USD 20 million to USD 115million (575% increase), making the early recovery the third largest requirement part (14.8%) of the total requirement. Additionally, the UNDP's "Time-critical Early Recovery Support for Debris Removal and Solid Waste Management for Community Affected by Typhoon Haiyan was sourced from the CERF (USD 21million).[32]"

8.10 Conclusion

This study focused on the following key questions: how much money is required for response activities, how and who outlines the whole amount, and how this funding is financially sourced.

[29] OCHA, Central Emergency Response Fund Life-Saving Criteria, V. Activities Eligible for CERF Funding.

[30] op. cit., CERF Secretariat (2011).

[31] OCHA Financial Tracking Service. https://www.fts.unocha.org/appeals/669/flows?f%5B0%5D=destinationClusterIdName%3A4340%3AEarly%20Recovery. Flow ID # 181,415.

[32] OCHA Financial Tracking Service. https://www.fts.unocha.org/appeals/441/flows?f%5B0%5D=destinationClusterIdName%3A2612%3AEarly%20Recovery%20and%20Livelihood. Flow ID #110,732, PHI-13/ER/61283/R.

The FA provides the answer to the first question by providing information about the total sum of the required funding and the details of each particular response. During the development of the FA, a humanitarian response project plan by individual organizations is drafted in each cluster. The sub-total of all projects in a particular cluster is the total requirement of the cluster response. Likewise, the grand total of all clusters is presented as the total sum of the required funding for the response. Here, the RC/HC and HCT along with humanitarian clusters together with the host government play a primary role in conducting assessments, analyzing the situation, and launching the FA.

To answer the second question, humanitarian funding is sourced from donor countries, humanitarian organizations, business sectors, and so on, and this is triggered by appeals. Appeals such as FA and HRP are launched during the early phase of the disaster response. The appeals consist of relevant information with a detailed humanitarian situation analysis.

Considering this understanding, this study reviewed four recent humanitarian response cases in the Asia-Pacific, with a focus on the early recovery cluster, which is the essential part of the "nexus" between response and development. Overall, it was found that the earlier the response phase, the lesser was the attention paid to the early recovery element. The met/unmet balance did not show any significant common features in any of the four cases. The CERF is an emergency response funding tool, and in the past, it hesitated to utilize its financial function during early recovery since it was not deemed to be a "life-saving" activity. However, the usability has since been reviewed, and CERF now invests in early recovery projects, as shown by the example of the Sulawesi earthquake and tsunami response in 2018. As seen in most cases, debris management is an important project under early recovery. This practice may be more mainstreamed in the future, and early recovery along with its reinterpretation of life-saving may become a frequent function of CERF response. Although early recovery is done in the early phase of disaster response, it may not be seen as the most pressing need compared to food and shelter. Greater involvement with adequate funding should be key in order to follow the Build Back Better concept, and realize Priority 4 in the Sendai Framework for Action 2015–2030.

Acknowledgements This study was supported by the Collaborative Research Project of the International Research Institute of Disaster Science, Tohoku University.

References

Publications with Author name

Elle C et al (2021) Funding disasters: tracking global humanitarian and development funding for response to natural hazards

Institutional Author

Development Initiatives (2021) Global humanitarian report 2021
Flash appeal, emergency response plan for Vanuatu, Tropical Cyclone Pam, March–June 2015
Flash appeal for response to the Nepal Earthquake, April–July 2015
Flash appeal revision for Nepal Earthquake, April–September 2015
Global cluster for early recovery (2015b)
Global Cluster for Early Recovery (2016) Guidance note on inter-cluster early recovery
The Government of Japan, GDFRR, The World Bank GFDRR (2012) The Sendai report: managing disaster risks for a resilient future
OCHA (2013) Addendum, Philippines: Typhoon Haiyan Action Plan 22 November, 2013
OCHA (2018a) UNDAC Field Handbook, Initial response planning and Flash Appeal, Chapter L
OCHA (2018b) CERF handbook
OCHA (2018c) Central Sulawesi Earthquake Response Plan (October 2018c-December 2018c)
OCHA (2021) 01 Step by Step Guide: humanitarian programme cycle 2022
OCHA (n.d.), Central emergency response fund life-saving criteria
Sendai Framework for Disaster Risk Reduction 2015–2030 (2015)
United Nations General Assembly Resolution 46/182, Strengthening of the Coordination of Humanitarian Emergency Assistance of the United Nations, 78th Plenary Meeting, 19 December, 1991
United Nations General Assembly Resolution 72/279, Repositioning of the United Nations Development System in the Context of the Quadrennial Comprehensive Policy Review of Operational Activities for Development of the United Nations System, Seventy-Second Session, Agenda Item 24 (a), May 31, 2018
United Nations General Assembly Resolution 60/124, Strengthening of the Coordination of Emergency Humanitarian Assistance of the United Nations, Sixtieth Session, 15 December, 2005

Websites

Inter Agency Standing Committee, IASC Transformative Agenda. https://www.interagencystandingcommittee.org/iasc-transformative-agenda. Accessed 9 Jan 2022
OCHA Financial Tracking Service. https://www.fts.unocha.org. Accessed 9 Jan 2022

Chapter 9
Is the Post Disaster Needs Assessment Important to Investment in Disaster Risk Reduction and Climate Change Adaptation? Lessons from Indonesia and the Philippines Based on Literature Review

Yasuhito Jibiki, Dicky C. Pelupessy, Daisuke Sasaki, and Kanako Iuchi

Abstract This chapter analyzes whether a significant relationship can be found between Post Disaster Needs Assessments (PDNA) and investment for Disaster Risk Reduction (DRR) and Climate Change Adaptation (CCA). We collected and summarized research trends in the literature regarding the implementation of PDNA in Indonesia and the Philippines and investigated whether discussion leading to investment in DRR and CCA has been made. PDNA has been implemented in both Indonesia and the Philippines for approximately 10 years. A literature review of the relevant articles on the PDNA implementation was conducted aiming to synthesize the findings from previous studies. No clear empirical research findings were identified indicating that PDNA had important significance for investment in DRR and CRR in Indonesia and the Philippines. One reason for this was the difficulty in government agencies skillfully utilizing techniques to incorporate CCA elements into the challenges of PDNA, which highlights disaster contexts rather than climate change issues.

Keywords Post disaster needs assessment · Disaster risk reduction · Climate change adaptation · Indonesia · Philippines

Y. Jibiki (✉) · D. Sasaki · K. Iuchi
International Research Institute of Disaster Science (IRIDeS), Tohoku University, 468-1 Aoba, Aramaki, Aoba-ku, Sendai 980-8572, Miyagi, Japan
e-mail: jibiki.yasuhito@gmail.com

Y. Jibiki
Center for Integrated Disaster Information Research (CIDIR), The University of Tokyo, 7-3-1 Hongo, Bunkyo-ku 113-0033, Tokyo, Japan

D. C. Pelupessy
Faculty of Psychology, Universitas Indonesia, Kampus UI Depok, Depok 16424, West Java, Indonesia

M. Ishiwatari and D. Sasaki (eds.), *Financing Investment in Disaster Risk Reduction and Climate Change Adaptation*, Disaster Risk Reduction,
https://doi.org/10.1007/978-981-19-2924-3_9

9.1 Introduction

This study aimed to synthesize findings on the effect of Post Disaster Needs Assessments (PDNA) implementation in Indonesia and the Philippines through a literature review and to examine if PDNA implementation might have any effect on investments in disaster risk reduction (DRR) and climate change adaptation (CCA). PDNA is a standardized method for comprehensively assessing damage in multiple sectors, including infrastructure, housing, and ways of funding social services. The results of PDNAs are utilized in recovery and reconstruction efforts (Nagami 2018). Theoretically, PDNAs incorporate the "build back better" concept emphasized in the Sendai Framework for Disaster Risk Reduction (SFDRR). To prevent or mitigate the same damage from occurring in the future, the needs assessment contains elements described as "investment in the future" that are expected to be included in future planning and implementation. Recent trends in government policies on global issues suggest that elements related to investment in CCA may be included in PDNAs. This is because, it is believed that unlike geological hazards such as earthquakes and volcanic eruptions, climate hazards are phenomena (e.g., floods, landslides, and tide surges) that are liable to be affected by climate change. PDNA has been carried out in both Indonesia and the Philippines for approximately 10 years. The experience of these two counties may provide evidence on the relationship between PDNA implementation and investment in DRR and CCA.

This chapter consists of four sections. The first section introduces the PDNA guidelines and published assessment reports indicating investment in DRR and CCA. The second section explains the method of the literature review. The third section presents the results of the literature review. The findings of the selected literature are classified into four topics: (1) Cases of PDNA implementation; (2) Legal system; (3) Problems related to the PDNA implementation; (4) Methodological issues. These classifications are solely inductive, rather than relying on existing theoretical frameworks. The relationship between PDNA and investment in DRR and CAA is also investigated in the third section, which is the primary objective of this study. The final section summarizes the findings and presents a discussion and an agenda for future research.

9.1.1 Investment in DRR and CCA Written in PDNA Guidelines

The European Commission, the United Nations, and the World Bank jointly published the PDNA guidelines in 2013. The guidelines consist of Volumes A and B. While Volume A shows more general concepts and procedures, Volume B provides more technical information for sector (e.g., education, gender, and health) specific assessments (European Commission et al. 2013).

In terms of explanation on investment for DRR, Volume A clearly notes that the costs for the "build back better" concept should be included in PDNAs. Volume B, in the environment sector, uses an example of mangroves for emphasizing the importance of investment for the future. It states that mangrove restoration contributes to the preservation of the ecosystem, but at the same time, healthy ecosystems could be utilized for defense against natural hazards (GFDRR et al. 2013). The statement "Investing in enhancing ecosystems for disaster risk reduction is a no-regret strategy […]" indicates that Volume B takes into account the importance of investments for DRR and environment issues, although it does not clearly use the term "climate change."

9.1.2 Investment in DRR and CCA Written in PDNA Reports

This section reviews whether published PDNA reports include aspects of investment for DRR and CCA.

According to the published needs assessment reports in Indonesia, while the needs assessment reports of the Indian Ocean Tsunami, the Yogyakarta earthquake, and the West Sumatra earthquake mentioned the importance of investment in DRR, no substantial evidence was found in the reports that indicated any items on investment were surely included in the budget for recovery and reconstruction. The needs assessment reports of the Indian Ocean Tsunami (National Development Planning Agency (BAPPENAS) Indonesia, 2005) and the West Sumatra earthquake (National Disaster Management Agency (Badan Nasional Penanggulangan Bencana [BNPB]) et al., 2009) clearly note importance investing in the enhancement of building safety against future earthquakes. The assessment report on the Yogyakarta earthquake (BAPPENAS et al., 2006) stated that investment in restoring irrigation infrastructure leads to mitigation of economic impact when facing future disasters. These three assessment reports did not mention the importance of investment in CCA.

Among the published needs assessment reports in the Philippines, the ones on the typhoon Sendong and the Bohol earthquake covered the aspect of investment in DRR and CCA to an extent. In contrast, the Ondoy and Pepeng, Pablo, and Yolanda reports did not clearly show specific evidence. In the assessment report of the typhoon Ondoy and Pepeng (Government of the Philippines et al., no date), while ideas of investment for the anticipated hazards and the "build back better" were mentioned, it did not show any concrete evidence for financial arrangements, such as calculation of the required budget. The assessment report of the typhoon Sendong (Office of Civil Defense and Regional Disaster Risk Reduction and Management Council—Region X, 2012) explicitly indicated that the aspect of "climate change resilience" was included in the recovery and reconstruction needs of agriculture. The assessment report for the typhoon Pablo (Davao Oriental Province 2013) recommended "climate change and disaster risk resiliency." However, it was not found if such discourse was reflected in the budget calculation for practical efforts. In the assessment report of the typhoon Yolanda (National Disaster Risk Reduction and Management Council,

2014), the health sector organized its needs based on the "build back better" principle. While investment was mentioned in the context of strengthening capacity building for dissemination of early warnings, it was not identified if the investment was really calculated on a monetary basis. The assessment report of the Bohol earthquake stated, "the Bohol Rehabilitation Plan took into account disaster-resistant and climate-resilient measures" (Province of Bohol, no date).

Jeggle and Boggero (2018) selected 14 PDNAs out of 55 that were conducted as of mid-2017 and assessed lessons from them. The 14 cases were from Bosnia and Herzegovina, Serbia, Mozambique, Vanuatu, Nepal, Georgia, Myanmar, Fiji, the Marshall Islands, Seychelles, Malawi, Ecuador, Sri Lanka, and Haiti. Jeggle and Boggero (2018) were jointly published by the European Commission, GFDRR, UNDP, and the World Bank, which were the founders of the PDNA guidelines. Thus, the report can be considered as an official and representative assessment of the latest progress on PDNA implementation. Regarding investment for DRR and CCA, Jeggle and Boggero (2018) noted as follows:

- "Several PDNAs refer to the recurrent nature of specified disasters, and nearly all mention 'the importance of climate change' in some respects." (Page 20)
- "Contemporary development concepts are embedded in PDNAs in such functions as measuring human development impacts, 'building back better' for resilient communities [...]. There is a continuing need to realize these features in an assessment process that goes beyond advocacy." (Page 42)
- "Additional considerations should be given to expand the scope of the PDNAs to address the impacts of climate change and its links to disaster risks [...]." (Page 42).

Jeggle and Boggero (2018) state proper investment in DRR or CCA has not been achieved, even though its importance was acknowledged in the PDNA implementation.

9.2 Methods

This study was an attempt to comprehensively collect the findings of previous studies through a literature review. We used Google Scholar to conduct our search as this tool identifies relevant literature in the form of reports, books, master's theses, and doctoral dissertations in addition to academic papers. The search terms were combinations of "the Philippines," "PDNA," and "Post Disaster Needs Assessment." The search was conducted on April 12, 2021, and resulted in 324 hits. Some of the articles used "PDNA" in contexts that were clearly different from "the Philippines and PDNA," thus, these were excluded. We identified 41 articles related to PDNA in the Philippines.

Of these 41 articles, 30 simply presented the results of PDNA (e.g., amount of damage in monetary terms). The remaining 11 contained more detailed treatments of cases in which PDNA was carried out and the associated problems. In the next section, we describe the results of our literature review focusing on these 11 articles.

As the details of our literature review regarding Indonesia were described in Jibiki et al. (2020), we provide only an overview here. The articles were collected using the same method as was used in the case of the Philippines. Ultimately, a detailed analysis was carried out on 25 articles.

9.3 Results

9.3.1 Cases of PDNA Implementation

Jibiki et al. (2020) revealed that many of the articles related to Indonesia were on the topic of the Indian Ocean Earthquake and Tsunami (2004) and the Yogyakarta earthquake (2006). Table 9.1 demonstrates more details about the disaster list in which PDNA has been implemented in Indonesia.

The results of the literature review of disasters in the Philippines showed that the cases in which PDNA was implemented were the 2009 typhoons Ondoy (international name: Ketsana) and Pepeng (international name: Parma), the 2011 typhoon Sendong (international name: Washi), the 2012 typhoon Pablo (international name: Bopha), and the 2013 typhoon Yolanda (international name: Haiyan) (Florano 2014; Florano et al. 2014; Gabriel et al. 2015; Gibb 2016; Villacin 2017; Florano 2018; Asia-Pacific Network for Global Change Research [APN] 2019; Kaneko 2015). Although one study indicates that PDNA was implemented for the Bohol earthquake (which occurred in October 2013; Domingo 2016), suggesting that a "Damage Assessment and Needs Analysis" was listed among the official documents (Province of Bohol,

Table 9.1 PDNAs conducted in Indonesia and the Philippines based on our literature review

Indonesia	The Philippines
• Indian Ocean Tsunami in 2004 • Central Java Earthquake in 2006 (also known as the Yogyakarta Earthquake) • West Sumatra Earthquake in 2009 (also known as the Padang Earthquake) • Mt. Merapi eruption in 2010 • Mt. Sinabung eruption in 2010 • Mt. Kelud eruption in 2014 • Lombok Earthquake (Nusa Tenggara Barat) in 2018 • Central Sulawesi disaster (earthquake, tsunami, and liquefaction) in 2018 • Series of floods in East Java (2000s) and in Jakarta (2000s)	• Typhoons Ondoy (international name: Ketsana) and Pepeng (international name: Parma) in 2009 • Typhoon Sendong (international name: Washi) in 2011 • Typhoon Pablo (international name: Bopha) in 2012 • Typhoon Yolanda (international name: Haiyan) in 2013 • Bohol earthquake in 2013

Note The disaster list in Indonesia was cited from Jibiki et al. (2020)

no date), this PDNA case is not included in the report issued by the World Bank (2020).

The search of the existing literature identified that Indonesia and the Philippines had started PDNA implementation prior to the establishment of the PDNA guidelines in 2013.

9.3.2 Legal System

In Indonesia, there have been studies (Jibiki et al. 2020) linking PDNA to the developmental plans of local municipalities (Hadi 2019) and budgetary issues (Nehru 2017; Gerungan 2020).

In the Philippines, PDNA was stipulated and carried out under the Disaster Risk Reduction and Management Act (Republic Act 10,121: Florano et al. 2014; Gabriel et al. 2015; Perwaiz and Sinha 2018). This refers to the existence of the Memorandum Circular No. 45 s2017, which lays out detailed rules for implementing PDNA (Villacin 2017; Qian et al. 2020). A memorandum in 2011 related to the link between DRR and CCA is also mentioned (APN 2019). This memorandum is known as the "Memorandum of Understanding between National Disaster Risk Reduction and Management Council and Climate Change Commission for Collaboration Programme on Philippine Climate Risk Reduction (the 2011 MOU)." It was signed by the National Disaster Risk Reduction and Management Council (NDRRMC) and the Climate Change Commission (CCC). The fact that this type of document existed prior to 2015, when the SFDRR was established, is highly significant in itself. The literature review did not reveal any clear mention of the importance of investment in disaster risk reduction.

9.3.3 Problems Related to Implementation

A study of problems related to the implementation of PDNA in the Philippines points out improvements in the skills of related personnel and cooperation among a wide variety of actors (Gabriel et al. 2015). Attention has also been paid to the relationship between the Office of Civil Defense (OCD), which leads PDNA efforts, and the National Economic Development Authority (NEDA), which is in charge of recovery and reconstruction under the NDRRMC. PDNA was carried out in the wake of the Sendong typhoon (international name: Washi) in 2011, but, though the NEDA drew up a document known as the "Strategic Action Plan for the Rehabilitation and Recovery of the Areas Affected by Tropical Storm Sendong (Washi)," confusion ensued as the details of that plan and the PDNA did not match (Florano et al. 2014; Gibb 2016). After typhoon Yolanda (international name: Haiyan) in 2013, it took approximately six months to compile the PDNA, but, unable to wait that long, the NEDA created its Reconstruction Assistance on Yolanda (RAY) before the PDNA report came out, which also caused confusion (Florano 2014; Ravago et al. 2017). The Source of the confusion was "the difference between RAY of NEDA and the PDNA of OCD which seems to have similar contents and methodologies, except for the analytical parts" (Florano 2014). Ravago et al. (2017) pointed out that the confusion in this context emerged because the preparation time for documentation was apparently different between the PDNA of OCD (six months) and the RAY of NEDA (the first version of RAY was unveiled in the next month after the landfall of the typhoon).

There has also been discussion on the issue of reviewing the contents of the PDNA. Kaneko (2015) proposes that the contents be examined, stating that "PDNA is described as outwardly being an objective sum of costs based on the estimated total cost of damage […], but in fact it is a municipal-level petition for financial resources for the purpose of development that is piggybacked on the 'Build Back Better' slogan advocated by the World Bank."

Problems with improving the skills of those working in bureaus charged with carrying out PDNA in Indonesia have also been pointed out (Jibiki et al. 2020). Problems with the Badan Penanggulangan Bencana Daerah (BPBD: Regional Disaster Management Agency) have been pointed out in particular (Nehru 2017; Sari 2017; Istiqomah 2019). Our literature review of Indonesia did not reveal findings indicating problems similar to those that exist between the OCD and the NEDA (in Indonesia, the National Disaster Management Agency [Badan Nasional Penanggulangan Bencana; BNPB] and the National Development Planning Agency have a similar relationship), indicating that this issue is unique to the Philippines.

9.3.4 Methodology

PDNA was refined based on guidelines created by the United Nations Economic Commission for Latin America (ECLAC). However, it is impossible to simply apply their methodology to all countries without any problems, and the problems with attempting to apply it in Indonesia have been pointed out (Jibiki et al. 2020). Daniell and Wenzel (2014) and Sharma (2010) argued almost the same point: the ECLAC methodology has unclear definitions for damage and loss calculation. Shafi (2011) noted that the level of physical damage and losses was overestimated in the Aceh Province in the Indian Ocean Tsunami, in which the ECLAC methodology was used. Medina and Santoso (2019) used a quantitative approach to the Damage and Loss Assessment method to find out the economic loss rate of the community in flood cases. In contrast, our literature review did not produce any findings in which this issue was brought up for the Philippines.

Our literature review of the Philippines revealed one work that emphasized the importance of gender analysis in PDNA (Perwaiz and Sinha 2018); our literature review of Indonesia also revealed one work on this issue (Harrowell and Özerdem 2019).

9.3.5 *PDNA and Investment in DRR and CCA*

Our literature review of the Philippines revealed three articles that discussed the implementation of PDNA in the context of DRR and CCA (Gabriel et al. 2015; Florano 2018; APN 2019). All three bring up the problem of integrating the DRR and CCA portions of PDNA: Gabrial et al. (2015) and Florano (2018) similarly pointed out some aspects of the integration problem, such as "(a) poor execution of existing policies, (b) unclear delegation of tasks, and (c) individualized implementation of harmonized CCA and DRR plans and projects" (Gabrial et al. 2015). APN (2019) revealed that the 2011 MOU did not induce any concrete advancement in terms of joint action between NDRRMC and CCC. We identified no findings or remarks related to the issue of any effect of PDNA on investment in DRR or CCA in this study of the Philippines literature review.

Our literature review of Indonesia revealed two articles (Sari 2017; Hinzpeter and Sandholz 2018) that discussed the implementation of PDNA in the context of DRR and CCA. Sari (2017) only pointed out that floods occurring in Indonesia as a result of climate change are increasing. Hinzpeter and Sandholz (2018) mentioned that the PDNA implemented in the wake of the Western Sumatra Earthquake, which occurred in 2019, included the following items: "Capacity building for Eco-DRR" and "Combat environmental degradation that enhances disaster risk." As with the results of our literature review for the Philippines, we were unable to identify any findings or remarks on a clear effect of PDNA on investment in DRR or CCA in the articles on Indonesia.

9.4 Conclusion

9.4.1 *Summary*

We aimed to synthesize findings on the effect of the PDNA implementation in Indonesia and the Philippines and to examine if PDNA implementation might have any function on investments in DRR and CCA through literature review. As introduced in Sect. 9.1.1, the PDNA guidelines indicated the importance of investment in DRR and CCA. While the published PDNA reports and Jeggle and Boggero (2018) acknowledged the importance of investment in DRR and CCA, they noted that such investment has not been fully brought into practice yet. Our literature review found that Indonesia and the Philippines had started PDNA implementation prior to the establishment of the PDNA guidelines in 2013. Regarding the hazard type, Indonesia mostly dealt with geological phenomena, while the Philippines mostly coped with typhoons. Both Indonesia and the Philippines institutionalized their own policies on PDNA. However, our literature review did not find clear indications of substantial investment in DRR and CCA in line with PDNA implementation. The PDNA implementation has been hampered by the limited capacity of relevant agencies in

Indonesia and the Philippines. In addition to limited capacity, the methodological difficulties were argued in Indonesia. The literature reviews conducted by the authors did not reveal any clear empirical research findings on the significance of PDNA for investment in DRR or CCA in either Indonesia or the Philippines. In other words, "mention" and "acknowledgement" of the importance of investment in DRR and CCA do not necessarily mean the substantial practice of investment. Our findings were consistent with implications in the published PDNA reports and Jeggle and Boggero (2018).

9.4.2 Discussions

In the Philippines, there are systemic problems with the OCD, which takes a leadership role in PDNA, and NDRRMC, for which the OCD serves as the executive office (Japan International Cooperation Agency 2017; Iuchi et al. 2019). This systemic problem has been generated by an issue of mixed policy arguments: there has been a long argument on the necessity for creating a new independent DRR agency or strengthening the existing OCD. As the Republic Act 10,128 had been planned to be evaluated since its endorsement, the so-called "sun-set review" of the Republic Act 10,128 has been begun since 2015 and relates to the possibility of re-structuring of OCD. Such backgrounds make the presence of OCD unstable for fully carrying out the PDNA implementation nationwide. There are also problems related to cooperation between the CCC, which has jurisdiction over the Climate Change Act, and the OCD and NDRRMC (Aryal and Dobson 2011; Iuchi et al. 2019). Even at the local government unit (LGU) level, it has been pointed out that it is not easy for LGUs to carry out DRR and CCA (Blanco 2015; Iuchi et al. 2019). Primarily, there is difficulty in expecting government agencies to skillfully utilize techniques to incorporate CCA elements into the challenges of PDNA, which highlight disaster contexts rather than climate change issues, immediately after being affected by the disaster.

In Indonesia, a variety of laws and ordinances are relevant (Jibiki et al. 2020). However, an examination of the documents related to PDNA implementation in the laws and ordinances of the Indonesian government indicates that there is no clear mention of investment in DRR or CCA. Discourse related to climate change in the National Disaster Management Agency (Badan Nasional Penanggulangan Bencana [BNPB]) (2008) does point out that the effects of climate change must be avoided, but this cannot be interpreted as implying that there is any clear promotion of investment. Although the term "climate change" itself is not used in either National Disaster Management Agency (Badan Nasional Penanggulangan Bencana [BNPB]) (2010) or National Disaster Management Agency (Badan Nasional Penanggulangan Bencana [BNPB]) (2011), the term "environment" is. "Environment" is an abstract term, but, based on its context, it can be understood to mean "environmental conservation" (protection of water resources, cessation of the destruction of forests, and preventing decreases in wildlife diversity). In the "Environment Sector" in National Disaster Management Agency (Badan Nasional Penanggulangan Bencana [BNPB]) (2011), there is mention of the estimated value of road reconstruction, but it is unclear if this includes an amount of money with elements of investment in the future. Specifically, it is impossible to interpret this as meaning investment in the future in anticipation of climate change.

The PDNA guidelines created by the World Bank and others were announced in 2013, but Indonesia and the Philippines had begun PDNA prior to the establishment of international standards. At the time PDNA was begun in these countries, however, organic ties between DRR and CCA had not yet come to practical realization. In addition, in Indonesia and the Philippines, PDNA implementation preceded the SFDRR, which clarified the very concept of investment in the future. The literature reviews of both countries did not reveal any elements of clear linkage between DRR and CCA or the investment in them.

9.4.3 Future Research

Although PDNA has been implemented in the wake of damage caused by the major typhoons that have struck since 2009, the fact that there have been few studies on this issue seems to be due to the limited amount of cases including empirical analyses of the investment in DRR and CCA. In contrast, according to the World Bank (2020), of the NDRRM Fund budget framework (total: PHP 105.7 billion) in the government budgets from the fiscal year 2015–2018, PHP 2.0 billion was allocated for "adaptation projects and activities of LGUs and community organizations (for the People's Survival Fund)." It is believed that these budget allocations were related to CCA, because the People's Survival Fund was established to support CCA activities. Although the memorandum of 2011 (see Sect. 9.3.1.2) was meaningless (APN 2019), the budgetary provision makes some type of activity (case) possible, and thus, there is a need to collect materials and conduct a detailed analysis in the future.

It was notable that many previous studies focusing on the Philippines did not mention laws and ordinances. A focus for future investigation is the fact that, other than national laws and memoranda, laws and ordinances such as executive orders, administrative orders, and joint memorandum circulars as well as all types of developmental plans include no elements related to PDNA or investment in DRR and CCA.

New discoveries may be made as a result of interviews with relevant people and the collection of further materials. For example, one future focus of study in the Philippines would be a literature review and onsite survey of the decision-making process of the CCC, which has jurisdiction over climate change laws. Additionally, given that the estimated value of recovery and restoration activities listed in the National Disaster Management Agency (Badan Nasional Penanggulangan Bencana [BNPB]) (2011) is being reviewed, there is a need to elucidate the details of this as well.

Acknowledgements We appreciate the valuable comments from Professor HIROTA Koki of Saitama University. This study was conducted as part of the JSPS KAKENHI Grant Number 20K12340. This work was partially supported by the collaborative research project of the International Research Institute of Disaster Science (IRIDeS), Tohoku University.

References

Aryal KR, Dobson OM (2011) Guidelines for integrating weather related disasters risk reduction and adaptation in Asia and the Pacific regions for senior government officials (Volume Two: Operation). http://www.nrl.northumbria.ac.uk/id/eprint/9111/1/Guidelines_Volume_2_REVISED_FIRST_DRAFT_(NRL).pdf. Accessed 10 May 2021

Asia-Pacific Network for Global Change Research (APN). (2019). Assessing the linkages between climate change adaptation, disaster risk reduction, and loss and damage: case studies in the low-lying coastal cities of Indonesia, Philippines, Thailand and Vietnam. Climate adaptation framework—Regional research final report. https://www.apn-gcr.org/wp-content/uploads/2020/09/747a07c1465c3a22c9e15289d5df567e.pdf. Accessed 10 May 2021

Blanco DV (2015) Disaster governance in the Philippines: issues, lessons learned, and future directions in the post-Yolanda Super Typhoon aftermath. Int J Public Adm 38(10):743–756

Daniell JE, Wenzel F (2104) The economics of earthquakes: a reanalysis of 1900–2013 historical losses and a new concept of capital loss vs. cost using the CATDAT Damaging Earthquakes Database. http://www.eaee.org/Media/Default/2ECCES/2ecces_eaee/1505.pdf. Accessed 17 Jan 2022

Davao Oriental Province (2013) Post disaster needs assessment for Typhoon "PABLO". https://www.recoveryplatform.org/assets/publication/PDNA/2009_Philippines_Typhoons_Ondoy_and_%20Pepeng.pdf. Accessed 4 Jan 2022

Domingo SN (2016) An assessment of the sectoral and institutional implementation of the NDRRMP. Philip Inst Dev Stud Discus Paper Ser 2016-49

European Commission, GFDRR, and the United Nations (2013) Post-Disaster Needs Assessment Guidelines, Volume A. https://www.gfdrr.org/sites/default/files/publication/pdna-guidelines-vol-a.pdf. Accessed 26 June 2022 GFDRR, European Commission, and the United Nations. (2013). PDNA GUIDELINES VOLUME B - Environment, https://www.preventionweb.net/media/75273/download. Accessed 26 June 2022

Florano ER (2014) Community governance for disaster recovery and resilience: Four case studies in the Philippines. Philip Inst Dev Stud Discus Paper Ser, 2014-38

Florano ER (2018) Integrated loss and damage–climate change adaptation–disaster risk reduction framework: the case of the Philippines. In: Zommers Z, Alverson K (eds) Resilience: the science of adaptation to climate change. Elsevier, Amsterdam, pp 317–326

Florano ER, Perez JS, Pinheiro ATK (2014) Building back a better nation: disaster rehabilitation and recovery in the Philippines. In: The XVIII International sociology association world congress of sociology, 13 July 2014. Yokahama, Japan, Pacifico Yokohama

Gabriel AV, Pulhin-Yoshida P, Lasco RD (2015) Assessing the linkages between CCA, DRR, and loss and damage in the Philippines. In: Proceedings of the resilient cities 2015 congress. https://www.researchgate.net/publication/307859198_Assessing_the_linkages_between_CCA_DRR_and_loss_and_damage_in_the_Philippines. Accessed 10 May 2021

Gerungan WM (2020) Penanggulangan Bencana Pada Tahap Pascabencana Menurut Undang-Undang Nomor 24 Tahun 2007 Tentang Penanggulangan Bencana (Disaster management in the post-disaster stage according to law number 24 of 2007 concerning disaster management). Lex Et Societatis 7(9):79–87

Gibb C (2016) Governmentality and exclusion in post-disaster spaces: Conducting the conduct of the survivors of typhoon Sendong in Cagayan de Oro, Philippines. Doctoral dissertation, Université de Montréal

Government of the Philippines, Asian Development Bank, the United Nations, World Bank, and Global Facility for Disaster Reduction and Recovery (n.d.) Typhoons Ondoy and Pepeng: post-disaster needs assessment main report. https://www.recoveryplatform.org/assets/publication/PDNA/2009_Philippines_Typhoons_Ondoy_and_%20Pepeng.pdf. Accessed 4 Jan 2022

Hadi S (2019) Learning from the legacy of post-disaster recovery in Indonesia for the acceleration of post-disaster recovery in Lombok. J Dev Plan 3(1):14–31

Harrowell E, Özerdem A (2019) Understanding the dilemmas of integrating post-disaster and post-conflict reconstruction initiatives: evidence from Nepal, Sri Lanka and Indonesia. Int J Disaster Risk Reduct 36:101092

Hinzpeter K, Sandholz S (2018) Squaring the circle? Integrating environment, infrastructure and risk reduction in post-disaster needs assessments. Int J Disaster Risk Reduct 32:113–124

Istiqomah U (2019) Disaster management—Studi Peran Pemerintah Daerah dalam Rehabilitasi dan Rekonstruksi Pasca Banjir di Kabupaten Sampang (Study of the role of local government in post-flood rehabilitation and reconstruction in Sampang District). Doctoral dissertation, University of Muhammadiyah Malang, Indonesia

Iuchi K, Jibiki Y, Solidum R Jr, Santiago R (2019) Natural hazards governance in the Philippines. Oxford University Press, Oxford research encyclopedia of natural hazard science

Japan International Cooperation Agency (2017) Data collection survey for strategy development of disaster risk reduction and management sector in the Republic of the Philippines final report. https://www.openjicareport.jica.go.jp/pdf/12285003.pdf. Accessed 10 May 2021.

Jeggle T, Boggero M (2018) Post-disaster needs assessment: lessons from a decade of experience. European Commission, GFDRR, UNDP, and the World Bank. https://www.openknowledge.worldbank.org/handle/10986/30945. Accessed 4 Jan 2022

Jibiki Y, Pelupessy D, Sasaki D, Iuchi K (2020) Implementation of post-disaster needs assessment in Indonesia: literature review. J Disaster Res 15(7):975–980

Kaneko Y (2015) Loss compensation and judicial access in the post-disaster recovery in Asia. J Int Cooperat Stud 22(2/3):1–42

Medina A, Santoso EB (2019) Peningkatan Resiliensi Ekonomi Masyarakat Berdasarkan Tingkat Kerugian Ekonomi di Kawasan Terdampak Kali Lamong Kabupaten Gresik (Increased Community Economic Resilience Based on the Level of Economic Losses in the Kali Lamong Affected Area in Gresik Regency). Jurnal Penataan Ruang 14(1):1–5

Nagami K (2018) Analysis on reconstruction policies and models of assistance for planning based on sector basis vulnerability evaluation: for achieving build back better. Doctoral dissertation, Toyo University

National Development Planning Agency (BAPPENAS) Indonesia (2005) Indonesia: preliminary damage and loss assessment. https://www.reliefweb.int/sites/reliefweb.int/files/resources/352868BF0EC9882885256F95005915EC-wb-idn-20jan.pdf. Accessed 4 Jan 2022

National Development Planning Agency (BAPPENAS), the Provincial and Local Governments of D.I. Yogyakarta, the Provincial and Local Governments of Central Java, International partners (2006) Preliminary damage and loss assessment yogyakarta and central java natural disaster. https://www.recoveryplatform.org/assets/publication/PDNA/CountryPDNAs/Indonesia_Yogjakarta%20Earthquake_2006_Preliminary%20Damage%20and%20Loss%20Assessment.pdf. Accessed 4 Jan 2022

National Disaster Management Agency (Badan Nasional Penanggulangan Bencana [BNPB]) (2008). Perka 11/2008: Peraturan Kepala Badan Nasional Penanggulangan Bencana Nomor 11 Tahun 2008 Tentang Pedoman Rehabilitasi dan Rekonstruksi Pasca Bencana (Perka 11/2008: Regulation of the head of the National Disaster Management Agency number 11 of 2008 concerning guidelines for post-disaster rehabilitation and reconstruction)

National Disaster Management Agency (Badan Nasional Penanggulangan Bencana [BNPB]) (2010). Perka 17/2010: Peraturan Kepala Badan Nasional Penanggulangan Bencana Nomor 17

Tahun 2010 Tentang Pedoman Umum Penyelenggaraan Rehabilitasi dan Rekonstruksi Pascabencana (Regulation of the head of the National Disaster Management Agency number 17 of 2010 concerning general guidelines for the implementation of post-disaster rehabilitation and reconstruction)

National Disaster Management Agency (Badan Nasional Penanggulangan Bencana [BNPB]) (2011). Perka 15/2011: Peraturan Kepala Badan Nasional Penanggulangan Bencana Nomor 15 Tahun 2011 Tentang Pedoman Pengkajian Kebutuhan Pasca Bencana (Regulation of the head of the National Disaster Management Agency number 15 of 2011 concerning guidelines for post-disaster needs assessment)

National Disaster Management Agency (Badan Nasional Penanggulangan Bencana [BNPB]), BAPPENAS, Provincial and District/City Governments of West Sumatra and Jambi and international partners (2009) West Sumatra and Jambi natural disasters: damage, loss and preliminary needs assessment. https://www.recoveryplatform.org/assets/publication/PDNA/2009_Indonesia_West_Sumatra_Earthquakes_DLNA.pdf. Accessed 4 Jan 2022

National Disaster Risk Reduction and Management Council (NDRRMC) (2014) Executive summary post-disaster needs assessment of TY Yolanda affected areas. https://www.recoveryplatform.org/assets/publication/PDNA/CountryPDNAs/Philippines_%20Typhoon_2014_Report.pdf. Accessed 4 Jan 2022

Nehru A (2017) Peran BPBD Kabupaten Malang Dalam Upaya Rehabilitasi dan Rekonstruksi Pasca Erupsi Gunung Kelud di Desa Pandansari Kecamatan Ngantang Kabupaten Malang (The role of BPBD [the local disaster management agency] of Malang Regency in efforts toward rehabilitation and reconstruction of post-Mt. Kelud eruption in Pandansari Village, Ngantang District, Malang Regency). Doctoral dissertation, University of Muhammadiyah Malang

Office of Civil Defense, Regional Disaster Risk Reduction and Management Council—Region X (2012) Tropical storm Sendong post disaster needs assessment final report. https://www.recoveryplatform.org/assets/publication/Final_PDNAReport_13June2012.pdf. Accessed 4 Jan 2022

Perwaiz A, Sinha S (2018) Post-disaster needs assessment for resilient recovery using space applications and innovative technologies. In: Shaw R, Shiwaku K, Izumi T (eds) Science and technology in disaster risk reduction in Asia. Academic Press, Cambridge, USA, pp 343–365

Province of Bohol. Post-Great Bohol Earthquake Rehabilitation Plan. http://www.jagna.gov.ph/wp-content/uploads/2015/01/BoholRehabPlan.pdf. Accessed 16 Aug 2021

Qian R, Signer BL, Skalon T, Marohombsar Z (2020) Public expenditure review: disaster response and rehabilitation in the Philippines. https://www.openknowledge.worldbank.org/bitstream/handle/10986/35064/Disaster-Response-and-Rehabilitation-in-the-Philippines-Public-Expenditure-Review-Technical-Report.pdf?sequence=1. Accessed 10 May 2021

Ravago MV, Roumasset J, Jandoc K (2017) Risk management and coping strategies: climate change and agriculture in the Philippines. In: The 9th ASAE international conference: transformation in agricultural and food economy in Asia, 11–13 January 2017, Bangkok, Thailand

Sari DA (2017) Peran Pemerintah Daerah Dalam Upaya Rehabilitasi dan Rekonstruksi Pasca Bencana Banjir di Kabupaten Sampang (Studi Kasus di BPBD Kabupaten Sampang) (Role of local government in rehabilitation and reconstruction efforts of flood disasters in Sampang District [Case study in Sampang Regency BPBD]. Doctoral dissertation, University of Muhammadiyah Malang

Shafi MS (2011) Providing sustainable health and education services in disaster recovery programs. In: Proceedings of the world reconstruction conference, pp. 299–320

Sharma SK (2010) Socio-economic aspects of disaster's impact: an assessment of databases and methodologies. Economic growth centre working paper series, Working paper no: 2010/01., Nanyang Technological University

Villacin DT (2017) A review of Philippine government disaster financing for recovery and reconstruction. Philippine Institute for Development Studies Discussion Paper Series. No 201721

World Bank (2020) Technical report public expenditure review: disaster response and rehabilitation in the Philippines. https://www.openknowledge.worldbank.org/bitstream/handle/10986/35064/Disaster-Response-and-Rehabilitation-in-the-Philippines-Public-Expenditure-Review-Technical-Report.pdf?sequence=1. Accessed 16 Aug 2021

Part III
Policy for Bridging Gaps

Chapter 10
Estimating the Benefits of Flood Protection—Reviewing the Transition of the Cost–Benefit Analysis Manual in Japan and the Implementation Status of the International Flood Protection Project

Masashi Sakamoto, Daisuke Sasaki, and Mikio Ishiwatari

Abstract Cost–benefit analysis is crucial for prioritizing and justifying disaster risk reduction projects. This study aimed to identify important issues in the application of various cost–benefit analysis methods. It reviewed the diverse evolving processes of these methods in Japan and examined their advantages and limitations. In Japan, cost–benefit analysis methods are commonly applied to investments in flood protection engineering projects to evaluate their benefits. Furthermore, the analysis also estimated the damage prevention benefits of river improvement and the river water level reduction benefits of dam construction. The assets in danger of being damaged included houses, factories, infrastructures, and short-term economic activities. However, estimating all the benefits generated by flood protection projects remains a challenge. For example, the benefits of saving human lives or mitigating long-term indirect costs have not been counted. Therefore, this study proposes some concrete directions in this regard for future studies.

keywords Cost–benefit analysis · International Flood Protection Project · Manual

M. Sakamoto (✉) · D. Sasaki
International and Regional Cooperation Office, International Research Institute of Disaster Science (IRIDeS), Tohoku University, Sendai, Miyagi, Japan
e-mail: masashi.sakamoto@tk.pacific.co.jp

M. Sakamoto
Disaster Prevention Planning Dept, Land Infrastructure Div, Pacific Consultants Co., Ltd, Tokyo, Japan

M. Ishiwatari
Graduate School of Frontier Sciences, The University of Tokyo, Chiba, Japan

M. Ishiwatari and D. Sasaki (eds.), *Financing Investment in Disaster Risk Reduction and Climate Change Adaptation*, Disaster Risk Reduction,
https://doi.org/10.1007/978-981-19-2924-3_10

10.1 Introduction

Disaster-related damages have been increasing because of urban development and climate change (Golnaraghl et al. 2020; Ward et al. 2017). Particularly, urban areas are often more likely to develop in coastal areas and are also more vulnerable to climate change because of their flat topography (Aerts et al. 2014). It has been estimated that flooding-related economic losses in 120 major mega cities around the world, such as New Orleans in the US, Guangzhou in China, and Bangkok in Thailand, among others, will range from US$2.5tn to US$3tn by 2050 (Abadie et al. 2017).

Investing in disaster risk reduction (DRR) is crucial for mitigating damages in cities (Jha and Stanton-Geddes 2013). The Sendai Framework for Disaster Risk Reduction at the Third United Nations World Conference on Disaster Risk Reduction in 2015 also emphasized the importance of investment for reducing disaster-related risks and damages (United Nations Office for Disaster Risk Reduction [UNISDR] 2015).

In this context, it is essential to demonstrate the benefits of promoting DRR investment. Most countries tend to use their disaster budgets during disaster response situations or recovery phases and often have limited funds (Kawasaki and Rhyner 2018).

Cost–benefit analysis—a method for indicating the benefit of a given investment—is a project evaluation method that focuses on efficiency (Tatano and Takagi 2005). In the field of flood protection projects, this method aims to measure the economic benefits and the costs and benefits created through the construction of flood protection facilities (e.g., levees and dams). The benefits of developing flood protection facilities include protection from direct damage (e.g., loss of life) and indirect damage (e.g., loss of property). In addition, it could also provide certain benefits associated with increased land productivity and the mental peace of mind associated with improved flood protection safety (Ministry of Land, Infrastructure, Transport and Tourism 2020).

10.2 Objective

This study focused on cost–benefit analysis, a method that has been used to demonstrate the benefits of investments. In Japan, the manual for economic survey on flood protection (draft) (hereinafter, the manual) was formulated in 1999 as a quantitative method for estimating the total damage caused by floods. Since then, minor changes have been made to this document but the method for estimating flood damages has not been fundamentally revised. In a project evaluation, the investment benefit of the relevant project is evaluated through a cost–benefit analysis based on this manual. However, the currently recorded benefits are based on a damage reduction calculation that can be converted into accurate monetary values representing the flood protection projects. These form only a small portion of the total benefits of flood protection projects. To visualize the benefits of flood protection projects, it is necessary to improve the system for translating benefits into monetary terms; it is also important to quantitatively estimate any benefits that may be difficult to convert into monetary terms.

This study reviewed the various transitions in flood damage estimation methods in Japan to examine the advantages and limitations of applying cost–benefit analysis methods to flood protection projects. Furthermore, to clarify the actual implementation status of cost–benefit analysis related to flood protection projects overseas, this study reviewed the application of cost–benefit analysis in flood protection projects implemented by the Japan International Cooperation Agency (JICA). Additionally, this study also discusses the current status of implementation and the ideal future of the evaluation methods.

10.3 Japan's Investment in Flood Protection

This section reviews the trends and benefits of flood protection investment in Japan. Flood protection has a long history. Military commanders in the age of the civil war—from 1550 to 1600—tried to protect lands from flooding. For example, the Kamanashi River, which deserves to be called a rampaging river in the Fuji River basin in the western part of the Kofu Basin in Yamanashi Prefecture, was once plagued by repeated floods. The series of hydraulic facilities, which are said to have been devised and built by Shingen Takeda, are collectively called the Shingen Embankment for "Shingen-dsutsumi." It is said that this construction reduced the damage caused by floods and promoted the development of the Kofu Basin (Nemoto et al. 2011).

The practice of investing in flood protection has continued for more than a century in Japan as part of the efforts to increase flood resilience. With the advent of modernization in the nineteenth century, national-level flood protection projects began to be undertaken (Ranghieri and Ishiwatari 2014). For example, the 1900s saw the start of a flood protection project on the Tone River to protect the Tokyo metropolis from flooding (Takahashi and Okuma 1979). The Japanese government sought to incorporate modern technology into these projects by hiring Dutch engineers (Takahashi and Uitto 2004). Furthermore, the government has also developed laws, budgetary systems, and long-term plans for implementing flood protection projects (Ishiwatari and Surjan 2019). The flood protection budget formed 0.5 to 1.3% of the government revenue until 1910. However, these budgets were often limited, and it was difficult to secure funds for flood protection; this situation led to frequent flood damage during World War II (Takahashi 2009). Particularly, nationwide damage was caused by the "Makurazaki typhoon," which landed on Makurazaki in Kyusyu island on September 17, 1945, just after the World War II (Kawata et al. 1992). Subsequently, from 1946 to 1959, the economic damages from floods in a single year ranged from 1 to 10% of the national income. Furthermore, during this period, more than 1,000 people lost their lives to flooding every year (Ishiwatari 2019).

Because of these disasters, the Japanese government increased its flood protection budget. About 1% of the national income was allotted to flood protection measures from 1960 to 1990. In the mid-1990s, the annual benefits from protection investments were estimated to be around US$55bn (Tsukahara and Kachi 2016).

Compared with 1960, the area protected from flooding doubled in the 2000s, and the flood protection rate increased from 24% in the 1960s to 56% in the 2000s (Ministry of Land, Infrastructure, Transport and Tourism [MLIT] Yearly). Furthermore, the government has been promoting urban flood projects and greater investment in such projects. The MLIT estimates that the cost–benefit rate on investment for urban flood prevention is 3.3. An investment of US$34bn at 2001 prices has prevented US$112bn worth of damages (MLIT 2004).

10.4 Positioning Cost–Benefit Analysis

The MLIT calculates the benefits of flood protection projects based on their amounts of damage reduction. Furthermore, this amount of damage is estimated by surveying the number of damaged houses after a flooding incident. It is published annually as part of the flood damage statistics. The total amount of damage caused by flooding, which forms the basis for calculating the benefits in cost–benefit analysis, is a significant factor for determining investment efficiency before flood protection projects. The amount of damage is translated into appropriate monetary values with the aid of inundation simulation. This amount of damage is calculated based on the damage rate in accordance with the inundation depth set in the manual.

This section provides an overview of the manual as follows. The manual broadly classifies flood protection project-related benefits into the flow and stock benefits. Flow benefits focus on monetary flows (e.g., the payments associated with a project) and are unrelated to the amount of damage. Subsequently, the stock benefit indicates the benefit of preventing direct and indirect flood damages and the upgradation of land use that will occur in the future (Fig. 10.1).

Among these benefits, for the damage items in yellow, the damage rate from inundation depth was set to estimate the amount of damage, which was calculated based on the formula shown in Tables 10.1 and 10.2.

The figures associated with each benefit are linked to the calculation formula in Table 10.2.

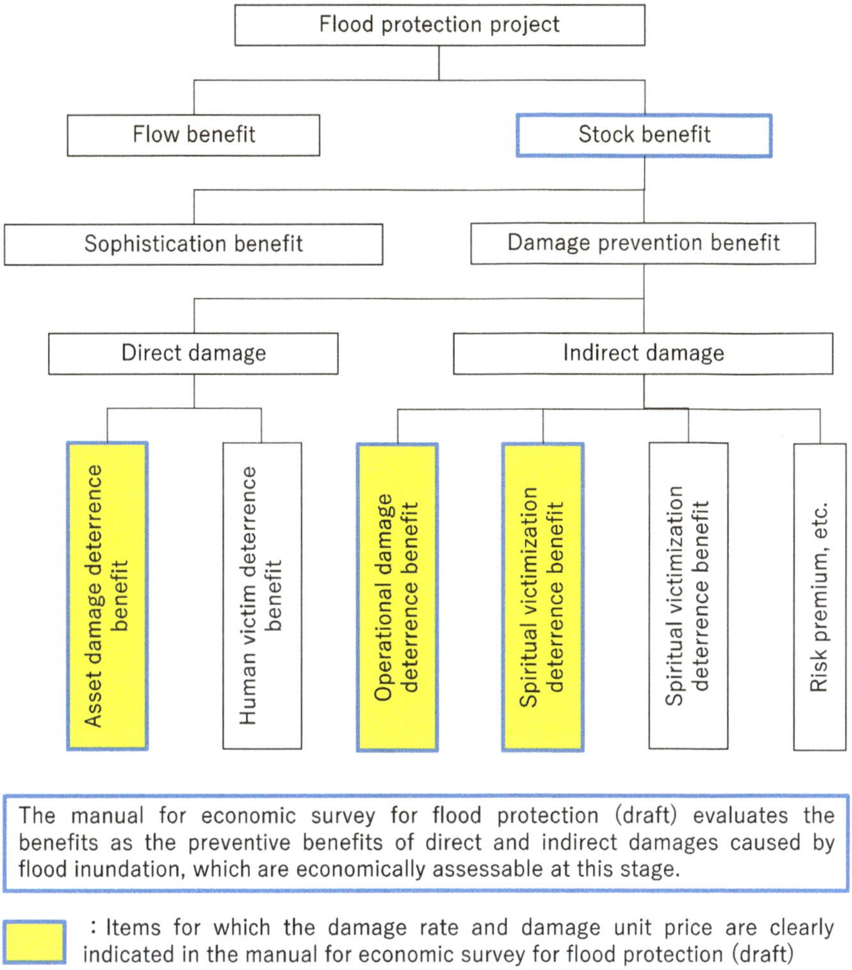

Fig. 10.1 Stock benefit of the flood protection project in the manual

10.5 Methodology

10.5.1 Reviewing the Revision Process of the Manual

Regarding the estimation method for assessing the amount of damage in flood protection projects, it is common to use the analysis outlined in the manual. However, it was found that the benefits considered in the cost–benefit analysis based on the manual only accounted for a portion of the total benefits of flood protection projects. Mainly, the items that could be translated into monetary terms were calculated; however, accounting for benefits that were difficult to convert into monetary terms posed

Table 10.1 Stock benefit of the flood protection project in the manual

Classification					Effect (damage)
Damage prevention effect (benefits)	Direct damage	Asset damage deterrence effect	General assets victimization	House	① Flood damage to residential and commercial buildings
				Household goods	② Flood damage to houses, automobiles, etc.
				Business depreciable assets	③ Flood damage to depreciable assets excluding land and buildings among fixed assets of business establishments
				Business inventory assets	③ Flood damage to business inventory
				Agricultural and fishermen's depreciable assets	④ Flood damage to depreciable assets excluding land and buildings among the fixed assets of farmers and fishermen involved in agriculture and fishery production
				Agricultural and fishermen's inventory assets	④ Flood damage to inventories of farmers and fishermen
			Agricultural product damage		⑤ Flood damage to crops
			Damage to public civil engineering facilities, etc		⑥ Flood damage to public civil engineering facilities, utility facilities, agricultural land, agricultural facilities
		Human victim deterrence effect			Life damage
	Indirect damage	Operational damage deterrence effect	Business stoppage	Household budget	Damage that hinders normal domestic work and leisure activities of flooded households
				Office	⑧ Suspension / stagnation of production at flooded business establishments (decrease in production)
				Public service	⑧ Suspension / stagnation of flooded public / public facility services
		Spiritual victimization deterrence effect	Cost of emergency measures	Household budget	⑦ Post-activities such as cleaning of flooded households, damages such as new expenses due to the purchase of alternatives such as drinking water
				Office	⑨ Damage similar to household budget
				National / local public organizations	Damage similar to households and interest and condolence money for urgent loans issued by municipalities, etc.
			Ripple effects of traffic disruptions	Roads, railroads, airports, ports, etc.	Ripple damage including surrounding areas due to traffic disruption such as roads and railroads
			Ripple effects of lifeline disconnection	Electricity, water, gas, communication, etc.	Ripple damage including surrounding areas due to suspension of supply of electricity, gas, water, etc.
			Business suspension ripple damage		Ripple damage including the surrounding area due to a decrease in production at surrounding business establishments due to a shortage of intermediate products and suspension of public and public services such as hospitals
		Spiritual victimization deterrence effect	Aspects associated with asset damage		Mental damage caused by property damage
			Aspects associated with operational damage		Mental impact due to operational damage
			Aspects associated with personal injury		Mental damage caused by personal injury
			Aspects associated with ex post facto damage		Mental damage caused by cleaning labor, etc.
			Aspects associated with ripple damage		Mental impact due to ripple damage
		Risk premium, etc.			Anxiety about the possibility of disaster
Sophistication effect					Rise in land prices due to improved hydraulic safety, etc.

Table 10.2 The damage amount calculation formula in the manual

①	House damage amount = Flooded house total floor area × Valuation per square meter of house × Inundation depth house damage rate
②	Household goods damage amount = Number of flooded households × Amount of household goods owned per household × Damage rate of household goods by inundation depth
③	Business asset damage amount = Number of employees at flooded establishments × (Valuation of depreciable assets per employee × Damage rate of depreciable assets at business establishments by inundation depth + Inventory asset valuation per employee × Inundation depth-based office inventory asset damage rate)
④	Amount of damage to farm and fishermen's assets = Number of flooded farms and fishermen × (Valuation of depreciable assets per unit × Agricultural and fishermen's depreciable asset damage rate by inundation depth + Inventory asset valuation per employee × Agricultural and fishermen's inventory asset damage rate by inundation depth)
⑤	Amount of damage to crops = Flooded paddy field/field area × Normal yield × Agricultural price × Crop damage rate by inundation depth
⑥	Damage to public civil engineering facilities, etc. = General asset damage amount × Ratio of damage to public civil engineering facilities to general asset damage
⑦	First aid costs at home = Number of flooded households × Cleaning days by inundation depth per household × Cleaning labor unit price per household per day + Number of affected households by flood depth × Alternative activity costs by inundation depth per household
⑧	Business suspension loss amount = Number of employees at flooded establishments × (Number of days of business suspension by flood depth + Number of days of stagnation/2) × Added value per employee by industry category
⑨	First aid costs at business establishments = Number of flooded establishments × Alternative activity cost per establishment by flood depth

a major issue (Tada 2014). The MLIT established the "Study Group on Evaluation Methods for River Projects" in September 2011. Since then, several discussions have been conducted to improve the accuracy of benefits estimates in the cost–benefit analysis and to quantify the damage items that have not been shaped into benefits. Therefore, this study aimed to review the revision process of the manual based on the documents of each study group and the summary of the discussion proceedings of the "Study Group on Evaluation Methods for River Projects." The details are explained in Sect. 10.6.2.

Adding to the methodology and revision of the manual on cost–benefit analysis, the psychological damage is also discussed as a challenge for calculating benefits that are difficult to convert into monetary terms. The details are discussed in Sect. 10.6.3.

10.5.2 The Implementation Status of Cost–Benefit Analysis for JICA Projects

As discussed previously, for flood protection projects in Japan, cost–benefit analysis is often conducted based on the manual. However, there is no uniformly established global-standard cost–benefit analysis method for implementing flood protection projects overseas.

Thus, this study sought to investigate the implementation status of cost–benefit analysis from the following unique viewpoint: the presence or absence of methodological differences (compared with the manual) for overseas flood protection projects by the JICA.

The JICA publishes project reports on its website (JICA Library). Two of the most recent reports on cost–benefit analysis implementations were extracted. In the following subsection, we confirm the contents of the project reports and investigate the implementation status of the cost–benefit analysis.

10.6 Examination

10.6.1 Changes in Flood Damage Estimation Methods

In Japan, the Annual Report of Statistics of the Public Works Bureau of the Ministry of Home Affairs (initiated in 1875) is said to be the oldest estimate of flood-related damage. After the war, knowledge regarding flood damage surveys was accumulated through the investigation of frequent floods. In 1961, the Ministry of Construction attempted to establish damage ratios for houses based on inundation depth and ground slope. The results of the flood damage surveys of 1961–1968 showed that the damage rate was revised annually. The results of these studies were integrated into the guideline on economic study on flood protection (hereinafter, the guideline), which was formulated in 1970 (Suetsugi 1998). In this outline, the house survey recommended the application of damage rate based on inundation depth to all household goods, business assets, and agricultural and fishing assets. The losses incurred due to business suspension and stagnation and the damages to public engineering facilities were calculated by multiplying other damages by a certain percentage. While making some assumptions, the outline formulated in 1970 established the method of setting the damage ratio based on inundation depth.

Furthermore, from 1998 onward, public works projects funded by the Ministry of Construction were required to conduct project evaluations including cost–benefit analyses. This requirement was established thanks to growing public criticism of public works projects and the fact that three ministries (i.e., the Ministries of Construction; Transport; and Agriculture, Forestry and Fisheries) had decided to introduce a project evaluation system from 1998 onward. Moreover, the Policy Evaluation Law was enacted in 2002, and project evaluation was incorporated as part of the policy evaluations based on this law. Against this backdrop, it became necessary to conduct a cost–benefit analysis for each public project that depended on or utilized government funds. The damage ratio based on inundation depth was revised based on the results of the flood damage surveys of 1993–1996. The manual was formulated in 1999. To conduct a cost–benefit analysis, the total benefit was calculated based on the annual average expected damage mitigation over a 50-year evaluation period, and the calculation method was specifically developed so that the benefits could be compared with the total cost by using a social discount rate of 4%. As described above, when the manual was updated, additional damage items were added, and the calculation method was standardized; however, the basic idea of damage ratio application based on inundation depth remained unchanged.

10.6.2 The Process of Revising the Manual for Economic Survey on Flood Protection (Draft)

The MLIT established a study group to examine the evaluation methods for river projects; accordingly, this group examined ways to enhance the evaluation methods for flood protection projects based on new knowledge, the latest data, and recent social conditions surrounding river projects (e.g., the impact of the Great East Japan Earthquake in this regard). The items to be considered by the study group were presented at the first study meeting in September 2011. The following four items were selected for consideration: (1) quantitative evaluation methods for evaluation items that had not been anticipated so far; (2) evaluation methods based on the fact that these methods were long-term utility assets; (3) treatment of cases where the investment efficiency of the remaining projects had fallen below the standard value; and (4) others (a reflection of the latest data and so on). Past large-scale floods have caused the loss of many lives, disruption of lifelines, and large-scale economic damage. The study group has been considering quantitative estimation methods for evaluation items that may be difficult to translate into monetary terms—especially if their benefits have not been recorded—but have a possible quantification. The MLIT has decided to formulate a "guidance for analysis of flood damage indicators (hereinafter, the guidance)" to quantitatively estimate assessment items that have not been accounted for with regard to the benefits of flood prevention. Furthermore, along with the cost–benefit analysis, the project will be comprehensively evaluated based on a benefit analysis using quantification indicators provided by the guidance (Fig. 10.2).

As a way forward, the group has set the damage rates based on the latest data. Furthermore, it has also examined whether the quantification indicators for damage indicator analysis could be transferred to the manual through monetary conversion.

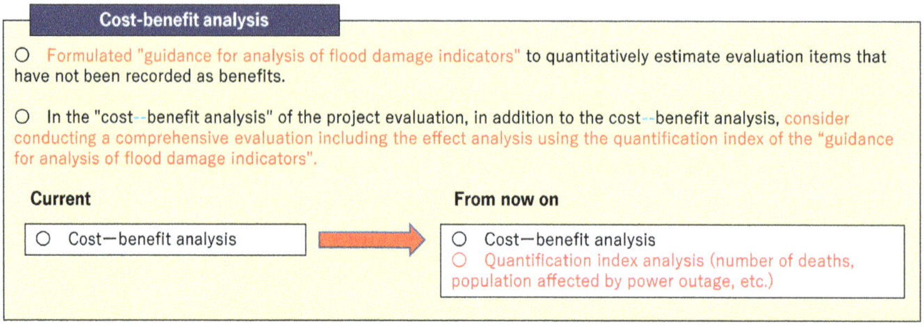

Fig. 10.2 Status of quantitative indicators calculated using the guidance (excerpt from the materials of the 3rd meeting on March 21, 2012 (MLIT 2012))

10.6.3 Calculating Psychological Damage

Flood damage is considered to cause physical as well as psychological and spiritual damage (e.g., anxiety) (Kuriki et al. 1996). However, a structure for assessing psychological damage and its quantitative assessment has not been established yet by the literature in this field.

These psychological damages tend to vary depending on inundation timing and other factors. One of the influencing factors included the victims' inability to envision the situation at the time of the disaster. Considering the future development of flood protection projects, special attention should be paid to not only the development of structural measures (e.g., levees and dams) but also non-structural measures (e.g., warning and hazard mapping). The implementation of warning and hazard mapping could result in decreased psychological damage, leading to a form of relief in flood protection contexts. Thus, it is essential to analyse and quantify the actual structure of the flood-related psychological damage; this could help to develop effective flood protection measures.

The psychological damage caused by the floods included the following: (1) shock from the loss of assets and household goods as well as worries about the future; (2) developing anxiety and fatigue as a result of having to live in evacuation centers (especially among the elderly); (3) developing fatigue because of post-flooding cleanup activities; (4) emergence of panic symptoms among children who have lost school textbooks, clothes, and other important items provided to them by family and friends; and (5) anxiety regarding the possibility of experiencing a flood situation again. Despite numerous indications to the contrary, to date, cost–benefit analyses in disaster reduction projects have not considered the benefits of reducing psychological damage partly because an appropriate economic evaluation methodology has not been considered.

The risk premium model study by Yokomatsu and Kobayashi (2000) examines the assumption of a risk-averse household utility function, which expresses households' avoidance preferences for risky situations. Therefore, this study used the risk premium model to express households' preferences as instantaneous consumption and the utility of physical assets. If a household buys a full-coverage insurance policy, the after the fact damage to the household will be assessed as zero, even if a disaster actually occurs. However, in practice, full-coverage insurance is often taken out on physical assets. Even if physical assets are restored, it can be expected that the affected households experience grief because of the loss of their memorabilia, fatigue from cleaning up after the disaster, and mental anguish because of their fear and anxiety about experiencing another disaster. Thus, victims are unable to avoid psychological damage.

The economic benefits of disaster reduction investment have been measured using reduction with regard to expected damage. Measurement of disaster reduction investment benefits based on expected damage constitutes just one hypothetical situation where asset damage can be fully covered through risk-fair disaster insurance (Tatano and Takagi 2005). However, no insurance market is available for hedging against psychological damage, and potentially affected households' exposure to psychological damage is often not manifested in the market. Therefore, to measure disaster reduction investment's benefits in reducing psychological damage, the existence of a hypothetical market must be assumed, and households' willingness to pay must be measured using the contingent valuation method (CVM). Wakigawa (2007) clarified a structure for the psychological damage suffered by affected households by using a covariance structure model. A methodology for measuring the psychological damage caused by flood protection projects as mitigation benefits (by using CVM) was also proposed.

10.7 Case Study

This section discusses the two most recent project reports involving cost–benefit analysis in flood protection projects to investigate their status. This study researched the status of cost–benefit analysis with a focus on methodological differences.

10.7.1 Chao Phraya River Basin Integrated Flood Management Plan, Thailand (2018)

The Chao Phraya River basin, which covers about one-third of Thailand's land area, has greatly benefited from the use of floods for cultivation. Rapid economic growth has led to the accumulation of assets and properties even in areas that face flood inundation risks; thus, flood risk reduction has emerged as an urgent issue here. Particularly, the 2011 floods submerged seven industrial estates in northern Bangkok, flooding 469 Japanese companies and causing economic damage worth 1.4 tn Thai baht. At the request of the Thai government, the JICA prepared the Chao Phraya River Basin Integrated Flood Management Plan (hereinafter, the JICA MP) in June

2013. Bangkok and its surrounding areas (east of the Tachin River and south of the Pasak River in Ayutthaya City) were selected by the Thai government as priority defense areas. The JICA MP proposed several structural and non-structural measures for flooding.

The Thai Royal Irrigation Department suggested that the construction of the Outer Ring Road spillway—as proposed by the JICA MP—should be considered as an integral part of the Third Outer Ring Road (east side), which was being promoted by the Expressway Bureau of the Ministry of Transport. This section presents the benefits of the Outer Ring Road spillway on the 2011 floods and its efficiency, along with assessing the benefits of the spillway project in terms of flood prevention benefits based on the manual (MLIT 2005).

In this project, the benefits to be evaluated were asset, operation, and the ex-post damage deterrence benefit (based on the manual). Thus, this project's estimated benefits strictly represented a portion of the benefits of the spillway project and could be considered as a modest benefit assessment. The Thai National Economic and Social Development Board (NESDB) and the Ministry of Industry estimated that the manufacturing and household sectors suffered 88.8% of the total direct damages caused by the 2011 floods. Regarding the indirect damage amount, the manufacturing and household sectors formed 66.8% of the total indirect damages. The total amounts of direct and indirect damages showed that they accounted for a majority (76.5%) of the total damage.

Therefore, for the manufacturing and household sectors, this project estimated the amount of assets to be considered as being in the target area from the latest statistical data and identified the amount of direct damage. The direct damages for the other sectors were estimated based on the ratio of the total direct damage of the manufacturing sector to the total direct damage of the household sector in the 2011 floods.

10.7.2 Medjerda River Integrated Watershed Management and Flood Protection Study, Republic of Tunisia (2013)

The Medjerda River flows through both Algeria and Tunisia. In 2003, its dam releases caused widespread floods. 6 people died, 27,000 people were evacuated, and the waterlogging period lasted for more than a month, resulting in damage to crops and houses as well as traffic blocks. Since then, similar floods caused by the releases of the Sidi Salem Dam in 2004, 2005, 2009, and 2012 have caused socio-economic damage and increased poverty, subsequently impeding the country's goal of achieving sustainable development. Subsequently, existing flood protection measures in the area are still at a relatively low level; thus, there is an urgent need for the formulation and implementation of comprehensive flood protection measures.

To improve this situation, at the request of the Tunisian government, the JICA conducted the "Integrated Water Management Planning Study for the Medjerda River Basin" (hereinafter, the development study) for 26 months (2006 to 2008). In this project, the investment efficiency of the structural remediation project was examined using the cost–benefit analysis method. The evaluation indices were benefit/cost ratio (B/C ratio) based on the cost and benefit of the difference between the cases where the project is implemented (with project) and the cases where it is not implemented (without project).

The specific calculation method for the benefits was as follows:

(1) Organize the assets in the assumed inundation area.
(2) Conduct flood inundation analysis for each probability of flood occurrence (2 to 10 years) and calculate the amount of flood damage in cases where the project has not been implemented.
(3) Calculate the amount of flood damage reduction based on the amount of flood damage when the project is implemented (no flood damage) and when project (2) is not implemented.

The items listed in the table below were set as indicators of direct and indirect damage (Table 10.3). This project calculated the amount of damage based on the manual.

10.8 Conclusions

In the Japanese context, significant progress was made on the manual through the MLIT's "Study Group on Evaluation Methods for River Projects," specifically, toward establishing the guidance as a method for quantitatively assessing damages that could not be easily translated into monetary terms. Quantitative estimation indices were presented for items that were considered to be quantifiable (e.g., human casualties and disruption of lifelines) during large-scale floods.

Future research directions should include the updation of damage ratios based on the latest flood data, current new benefit items, and economic evaluation. Regarding

Table 10.3 Direct and indirect damages

Damage classification	Nature of damage
Direct damage	· House damage · Damage to household items · Damage to business assets · Damage to farmers' property · Crop damage · Damage to public civil engineering facilities
Indirect damage	· Loss or suspension of business · First aid costs at home · First aid costs at business establishments

psychological damage, this issue is still in its early research stages because no specific evaluation method is described in the manual (Table 10.4).

This study examined several examples of cost–benefit analyses in overseas flood protection projects, where a series of economic analyses were conducted based on the manual used in Japan. However, only some of the benefits that could be estimated were extracted. Cost–benefit analyses are rarely conducted at the same level as in Japanese flood protection projects. This is because of the insufficiency of geographic information data available for estimating damage prevention benefits. Furthermore,

Table 10.4 Key topics and issues to be discussed in the future (excerpts from the materials of the 8th Study Group on January 9, 2020 (MLIT 2020))

Classification	Task	Issues
The updation of the damage rate	Accumulation of the latest flood data	• A flood damage fact-finding survey was essential even during the heavy rains of July 2018 and in the midst of Typhoon No. 19, which was the first typhoon of the year, to reflect the effects of the same in the damage rate updation process • It is important to consider a method for continuously and efficiently acquiring and accumulating data
	Review for a more probable calculation method	• It is necessary to continue considering the benefits of fluid strength on house damage and the various differences between business types and formats due to the damage caused to business establishment assets (depending on the accumulation of flood data) • Subsequently, in cases where it is possible to estimate the distribution of the damage rate based on the data accumulated so far and standard models (e.g., damage to household goods) and not only the accumulation of flood data, it is important to find a more reliable calculation method by reviewing the methods individually

(continued)

Table 10.4 (continued)

Classification	Task	Issues
New benefit items	Examination of items that can be newly translated into monetary terms	• The recording of benefits in the cost–benefit analysis is only a part of the various benefits of the hydraulic engineering project, and it is necessary to continue considering new items that can be converted into monetary terms
	Quantification of items that are difficult to convert into money	• It is necessary to (1) make quantitative estimates for items that cannot be recorded as benefits (e.g., economic ripple damage) and (2) to continue efforts toward comprehensive evaluation
Evaluation of economic efficiency	Evaluation based on the fact that a hydraulic facility is a long-term utility asset	• The socio-economic situation regarding the ideal evaluation method (social discount rate, treatment of the economic internal rate of return, etc.) is based on the fact that hydraulic facilities (e.g., embankments) are assets that will have long-term benefits if they are properly managed. It is necessary to proceed with the examination based on such factors
Others	The use of the capacity–benefit calculation method for maintaining the normal functioning of running water	• It is important to continue considering methods other than the alternative method (replacement construction cost) that is currently used

there are other issues including the unsuitability of data regarding damage rates based on inundation depth for local conditions.

To improve the accuracy of cost–benefit analyses, statistical data (e.g., the census and statistical census data), as well as geographic information data, have to be systematically developed in the future. Furthermore, it is necessary to define indicators so that they can be tailored to the social conditions of each country (e.g., damage rates based on inundation depth).

Cost–benefit analysis is a valuable tool used for project evaluation. Future studies should develop methods for analyzing the benefits from the entire basin based on the past accumulated investment.

Acknowledgements This study was supported by the Collaborative Research Project of the International Research Institute of Disaster Science, Tohoku University. This study was also partially supported by the Japan Society for the Promotion of Science KAKENHI, Grant Number 21H03680.

References

Abadie LM, Galarraga I, de Murieta ES (2017) Understanding risks in the light of uncertainty: low-probability, high-impact coastal events in cities. Environ Res Lett 12(1):014017

Aerts JC, Botzen WJ, Emanuel K, Lin N, de Moel H, Michel-Kerjan EO (2014) Climate adaptation. Evaluating flood resilience strategies for coastal megacities. Science 344(6183):473–475. https://doi.org/10.1126/science.1248222. PMID: 24786064

Golnaraghl M, Surmlnskl S, Kousky C (2020) Building flood resilience in a changing climate: insights from the United States, England and Germany. The Geneva Association, Zurich

Ishiwatari M (2019) Investing in disaster risk reduction: scale and effect of investment in flood protection in Asia. UN Office for Disaster Risk Reduction, Geneva

Ishiwatari M, Surjan A (2019) Good enough today is not enough tomorrow: challenges of increasing investments in disaster risk reduction and climate change adaptation. Prog Disaster Sci 1(1):100007

Jha AK, Stanton-Geddes Z (eds) (2013) Strong, safe, and resilient: a strategic policy guide for disaster risk management in East Asia and the Pacific. The World Bank, Washington, DC

Kawasaki A, Rhyner J (2018) Investing in disaster risk reduction for resilience: roles of science, technology, and education. J Disaster Res 13(7):1181–1186

Kawata Y, Onmae M, Oka T, Tsuchiya Y (1992) Recreation of typhoon disasters in postwar Japan (1) – Makurazaki typhoon. DPRI Ann 35(B-2):403–432

Kuriki M, Imamura Y, Kobayashi H (1996) Economic evaluation of intangible effects of flooding. J JSNDS 15:231–240

Ministry of Land, Infrastructure, Transport and Tourism (MLIT) (2004) Program evaluation on integrated flood management with river basin. MLIT, Tokyo (in Japanese)

Ministry of Land, Infrastructure, Transport and Tourism (MLIT) (2005) Manual for economic survey for flood protection. MLIT, Tokyo (in Japanese)

Ministry of Land, Infrastructure, Transport and Tourism (MLIT) (2012) The materials of the "study group on evaluation methods for river projects". In: 3rd meeting on 21st March 2012

Ministry of Land, Infrastructure, Transport and Tourism (MLIT) (2020) The materials of the "study group on evaluation methods for river projects". In: 8th meeting on 9th January 2020

Nemoto Y, Nakayama D, Matsuyama H (2011) Reevaluation of Shingen-tsutsumi based on inundation flow simulations with special focus on the flood control facilities along the Midai River. Geogr Rev Jpn Ser B 84(6):553–571

Ranghieri F, Ishiwatari M (2014) Learning from megadisasters: lessons from the Great East Japan Earthquake. The World Bank, Washington, DC

Suetsugi T (1998) History of technique and guideline for economic evaluation of flood damage and flood protection projects. Hist Stud Civ Eng 18:603–618

Tada N (2014) Research on the sophistication of flood damage estimation methods. Ph.D. thesis, Tokyo University

Takahashi Y (2009) History of water management in Japan from the end of world war II. Int J Water Resour Dev 25(4):547–553

Takahashi Y, Okuma T (1979) Historical review of floods and their damage in the tone river basin. Nat Disaster Sci 1(1):35–50

Takahashi Y, Uitto JI (2004) Evolution of river management in Japan: from focus on economic benefits to a comprehensive view. Glob Environ Change 14:63–70

Tatano Y, Takagi A (2005) Economic analysis of disaster prevention. Keiso Shobo

Tsukahara K, Kachi N (2016) Using data and statistics to explain investment effectiveness on flood protection. J Disaster Res 11(6):1238–1243

United Nations Office for Disaster Risk Reduction (UNISDR) (2015) Sendai framework for disaster risk reduction 2015–2030. UNISDR, Geneva

Wakigawa K (2007) Research on the sophistication of evaluation methods for hydraulic economic benefits. Ph.D. thesis, Kyoto University

Ward PJ, Jongman B, Aerts JCJH, Bates PD, Botzen WJW, Diaz Loaiza A et al (2017) A global framework for future costs and benefits of river-flood protection in urban areas. Nat Clim Change 7(9):642–646

Yokomatsu M, Kobayashi K (2000) Economic benefit of physical risk reduction by disaster prevention investment. Doboku Gakkai Ronbunshu 2000(660):111–123

Chapter 11
Nature-Based Solutions for River Restoration in Japan

Keigo Nakamura

Abstract This chapter describes the range of nature-based solutions being implemented in Japan, with a focus on river restoration. In Japan, modern nature-based solutions are often referred to as green infrastructure (GI). However, even before the term "GI" was coined, several nature-based disaster prevention policies had been effectively implemented, such as nature-oriented river management. Efforts toward GI have been expanding in Japan since the Great East Japan Earthquake of 2011. In recent years, GI has attracted further attention owing to the increased frequency and severity of climate change-induced natural disasters, and a public desire for well-being. Consequently, Japan's Ministry of Land, Infrastructure, Transport, and Tourism established the GI Public–Private Partnership Platform in 2020 to accelerate the implementation of GI by leveraging the knowledge and technologies of various stakeholders. With the emergence of GI policies, greater emphasis has been placed on public happiness (well-being), regional development, and urban planning. This chapter discusses GI developments in Japan, using case studies of river management as examples.

Keywords Green infrastructure · Nature-oriented river management · Watershed-based flood management

11.1 Introduction

According to the IPBES report (2019), an estimated one million species, including those yet undiscovered, are threatened with extinction, including 41% of the amphibians and 33% of the reef-building corals (IUCN 2021a). Alarmingly, Global Biodiversity Outlook 5 (GBO5), which reports on the status of global biodiversity, noted that none of the 20 Aichi Biodiversity Targets were met by their 2020 deadline (CBD 2019). To compound this problem, the intensification of natural disasters predicted

K. Nakamura (✉)
Japan Riverfront Research Center, Tokyo, Japan
e-mail: keigo.roppongi@gmail.com

M. Ishiwatari and D. Sasaki (eds.), *Financing Investment in Disaster Risk Reduction and Climate Change Adaptation*, Disaster Risk Reduction,
https://doi.org/10.1007/978-981-19-2924-3_11

to result from anthropogenic climate change has already become a reality. It is, therefore, necessary to not only mitigate the effect of these disasters but to also establish disaster prevention measures and resilient societies that are in harmony with nature. In this context, the key phrase "nature-based solutions" (NbS) has attracted international attention and featured as one of the themes at the UN Climate Action Summit 2019.

The IUCN defines NbS as "actions to protect, sustainably manage, and restore natural or modified ecosystems, that address societal challenges effectively and adaptively, simultaneously providing human well-being and biodiversity benefits" (IUCN 2021b). NbS is an umbrella term that encompasses several related concepts, such as Eco-DRR, an ecosystem-based approach to disaster prevention and mitigation; EbA, an ecosystem-based approach to climate change adaptation; and green infrastructure (GI), a form of infrastructure development that utilizes diverse natural functions. In Japan, the term GI (referred to as "green infra" in Japanese) is the most widespread NbS, and many Japanese-language books have been published on this subject. The term is also widely understood among infrastructure professionals, as 63% of the environmental department staff in cities (over 200,000 people) understand its meaning (Manabe 2021).

In recent years, the Ministry of Land, Infrastructure, Transport, and Tourism (MLIT) announced the "Green Infrastructure Promotion Strategy" (MLIT 2019), based on the policies of the Japanese government. To enable the expansion of this strategy, the "Green Infrastructure Public–Private Partnership Platform" (MLIT 2021a) was subsequently established. However, in Japan, efforts to conserve and restore the natural environment have been promoted since long before the terms NbS and GI were coined. A major example is the Nature-Oriented River Works (NORW; now called Nature-Oriented River Management [NORM]) initiated by the MLIT in 1990 (Nakamura et al. 2006). This chapter reviews the status of the GI initiatives in Japan, with a focus on river restoration as a practical example.

11.2 Initiatives for Green Infrastructure

11.2.1 Green Infrastructure in Japan, Europe, and USA

GI emerged in the late 1990s, mainly in Europe and the United States; however, its meaning differs from country to country.

The EU defines GI as "a strategically planned network of natural and semi-natural areas with other environmental features designed and managed to deliver a wide range of ecosystem services such as water purification, air quality, space for recreation, and climate mitigation and adaptation" (EU 2013). In Europe, there is a strongly held view that GI should be used to advance existing ecosystem network planning while contributing to disaster prevention, and boosting the economic benefits of ecosystem services.

However, in the United States, GI is specifically used to refer to sewer systems and stormwater measures that prevent sewage overflow during heavy rains. The Environmental Protection Agency defines GI as "the range of measures that use plant or soil systems, permeable pavement or other permeable surfaces or substrates, stormwater harvest and reuse, or landscaping to store, infiltrate, or evapotranspirate stormwater and reduce flows to sewer systems or to surface waters" (EPA 2019). However, it also states that these measures will provide additional benefits for flood control and environmental protection. In this respect, the focus on the multifunctionality of GI is similar to the values of the EU.

In Japan, GI came into focus following the Great East Japan Earthquake in 2011. From the beginning, the primary focus has been on disaster prevention using the functions of nature. For example, in addition to difficult countermeasures against powerful tsunamis, the concept of green seawalls using tree plantations has attracted attention as a GI solution. Additionally, considering the current situation of severe flooding due to climate change, GI is attracting further attention as a new concept for more resilient infrastructure.

The National Spatial Strategy (NSS; MLIT 2015a), a basic plan for the development of the national land in Japan, states that "GI is a concept that promotes the creation of sustainable and attractive national land and regions by utilizing the diverse functions of the natural environment (e.g., providing habitats for living organisms, creating favorable landscapes, and reducing temperature increases) in both hard and soft aspects of social infrastructure development and land use." Even within Japan, the perception of GI differs considerably from person to person, but it is often viewed simply as "nature-based infrastructure development."

As examined above, the concept of GI differs around the world, with the United States being focused on sewage systems (sewage type), and Europe being focused on ecology and conservation (ecology type). Japan's concept of GI is much broader than that of the United States, being relatively similar to that of Europe. However, as natural disasters represent a significant threat to the Japanese society, their primary focus for GI is disaster prevention (disaster prevention type).

11.2.2 Green Infrastructure in Japan

Nearly 70% of Japan's land area is covered with forests (Fig. 11.1), which is more than double that of Western Europe and the United States, and even more than that of Brazil, which is renowned for its Amazon rainforest, and similar to that of Scandinavian countries (The World Bank 2020). Furthermore, Japan's entire land area is classified as one of the world's 36 biodiversity hotspots (Conservation International 2021). Thus the value of Japanese ecosystems is extremely high. Therefore, Japan is a country with high potential for GI, and the formation of national infrastructure in harmony with nature through GI initiatives will not only create a sustainable country but will also be of high value internationally.

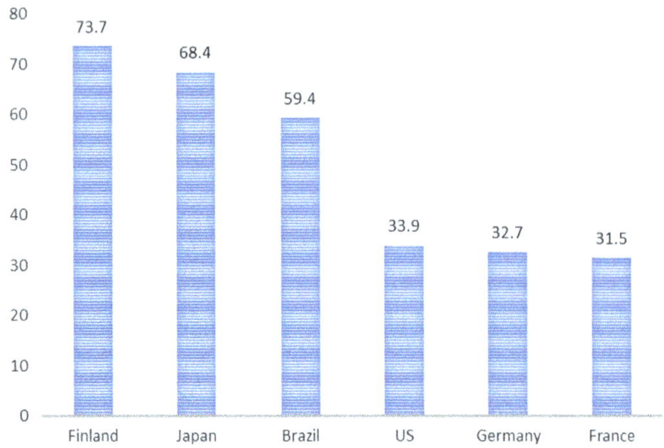

Fig. 11.1 Percentage of forest area (%) (The World Bank 2020)

As stated earlier, the devastating Great East Japan Earthquake of 2011 triggered a heightened focus on GI in Japan. In 2013, the Ecosystem Conservation Society of Japan organized a conference entitled "From Gray Infrastructure to Green Infrastructure." In 2014, the Science Council of Japan released a proposal titled "Recommendations for the Use of Ecosystem Infrastructures in Reconstruction and National Land Resilience" (using the analogous term "ecosystem infrastructure" instead of GI; Science Council of Japan 2014). Many from the academic community made proposals for reconstruction from the disaster based on GI.

In response to these developments, NSS (MLIT 2015a), a comprehensive plan to promote the use, development, and conservation of national land at the national level, officially included the promotion of GI initiatives as a policy that should be pushed forward by the government, and called for the "conservation, restoration, and utilization of the natural environment through the promotion of GI initiatives." It states that the concept of GI will be actively incorporated into the build-up of ecological networks, infrastructure development, and land use. In the same year, the MLIT also mentioned GI in its Fourth Priority Plan for Social Infrastructure Development (4th PPSID) (MLIT 2015b), which is designed to promote intensive and efficient social infrastructure development projects. Although the term GI appears in only four places in the 4th PPSID, it appears in 32 places in the 5th PPSID (MLIT 2021b), indicating its growing policy importance.

In accordance with these movements, MLIT formulated the "GI Promotion Strategy" in July 2019 (MLIT 2019), which includes three policies: (1) creating an environment for the mainstreaming of GI, (2) improving support for the promotion of GI, and (3) developing evaluation methods for GI. Based on Policy (1), the GI Public–Private Partnership Platform (GIPF) was established in March 2020 to promote collaborative efforts among citizens, industry, government, and academia.

GIPF, with MLIT as the secretariat, has three subcommittees under the Steering Committee, i.e., the Planning and Public Relations Subcommittee, the Technology Subcommittee, and the Finance Subcommittee, with both committee and general members participating in each subcommittee. The author is a member of the Steering Committee for the platform and the head of the Technology Subcommittee.

The platform has since begun full-scale activities, including holding a symposium in June 2020 and executive committee meetings for each subcommittee in July 2020. In August, the platform started accepting applications for the GI Grand Prix; 22 cases were selected for the Excellence Award from 117 cases nationwide, and a Grand Prize was awarded in four categories. To consolidate and share information on GI initiatives and technologies, a reference collection was created, and at the end of FY2020 (March 2021), three reference collections were published in Japanese: "GI Case Studies," "GI Technology Collection," and "GI Finance Collection." The number of members of the platform stands at 1,233 as of October 2021 and includes individuals and companies. As many of these memberships are held by companies and local governments, the actual number of participants in the platform is estimated to be much higher. Various activities using this platform have already begun, including a GI webinar, working groups on specific topics, and project matching between local government and companies. Thus we expect to see the expansion of many GI technologies.

11.2.3 Green Infrastructure and Finance

One of the characteristics of GI is that it emphasizes not only disaster prevention and the environment but also the economy. In the development of GI, it is important to proactively adopt diverse financing methods, such as methods of combining public and private funds, that have not been widely used by conventional infrastructure managers mainly depending on public funds.

The Private Finance Initiative (PFI) model is currently being used in the development of waterfront areas and parks. Recently, prefectural governments have issued tens of billions of yen in green bonds to encourage prefectural residents to invest in environmental projects, thereby promoting public understanding and ownership of projects.

It is also necessary to pay attention to Payments for Ecosystem Services (PES) (e.g., a tax paid by downstream citizens to conserve upstream forest) as a fiscal policy tool to promote funding for environmental conservation and to contribute to both conservation and regional development.

A social impact bond (SIB) is another interesting financing method in which unique public–private partnerships fund effective social services through performance-based contracts. The Sampo-Yoshi Fund of Higashi-Ohmi City has initiated a small-scale river restoration project under a performance-based private sector contracting system (Pay for Success and Social Impact Bonds [PFS/SIB]).

The success or failure of this project will be closely monitored to assess its viability as a potential method for financing river restoration in the future.

In July 2020, MLIT received a report from the Panel on Infrastructure Development announcing a shift to Watershed-based Flood Management (WFM) for flood prevention amidst climate change (2020). The "utilization of GI" is also mentioned in the report. Though MLIT has the authority and public budget for river management under the River Law, it has limited authority and budget for watershed management. Therefore, to implement the WFM approach, it is necessary to secure a funding system and promote the use of diverse financing strategies like those mentioned above.

11.3 Practice of Nature-Based Solutions and Green Infrastructure—Case Studies of River Restoration

11.3.1 Tashizen River Management (Nature-Oriented River Management [NORM])

In Japan, the Tashizen River Works, or NORW, began in 1990, borne out of growing public environmental awareness (Nakamura et al. 2006; Waley and Aberg 2011). This policy primarily aimed to conserve beautiful landscapes and natural habitats, and flood prevention works have been implemented with these aims in mind. By the early 2000s, NORW had been successfully implemented in tens of thousands of locations and had convinced many skeptical river managers of the value of NbS. In 1997, the spirit of NORW was extrapolated to the River Law, the basic law for river management in Japan, and the phrase "improvement and conservation of river environment" was listed as an objective in Article 1 of the River Law. Although this policy was a great success in terms of raising awareness of the river environment among river administrators and launching nature restoration projects, it was not always easy to understand nature and practice nature-conscious river restoration, and the number of problematic restoration cases gradually increased. Therefore, in 2006, a new basic principle was created, and the name was slightly revised from NORW to Nature-Oriented River Management (NORM; MLIT 2006). The definition also broadened to mean "river management that conserves and creates the habitat, growth, and breeding environment of living organisms and diverse river landscapes inherent to rivers, while taking into account the natural activities of the river as a whole and considering harmony with local life, history, and culture," whereas the purpose of NORW was simply to conserve and create beautiful landscapes and natural habitats.

In addition, although it was previously the policy to not prepare manuals in order to avoid uniform river design, a set of technical standards for river restoration for small- and medium-sized rivers was prepared in 2008 (and revised in 2010) based on improved knowledge of basic matters (MLIT 2010). In 2011, the Point Book III for NORM was prepared (Research Group on Nature-oriented River Management

Fig. 11.2 Artificial river before restoration (April 26, 2007) (Kamisaigo River, photo by Dr. Hironori Hayashi)

2011), which explains the technical standards in an easy-to-understand manner. Since then, understanding of the technical aspects of NORM has increased, the practice of NORM has spread, and the technical level of river restoration has improved.

Here, we introduce the case of the Kamisaigo River (Fukutsu City, Fukuoka Prefecture) as an example of a successful NORM implementation (Hayashi et al. 2016). The Kamisaigo River is an urban river that runs through a residential area of Fukutsu city. The river improvement project was conducted in conjunction with the land readjustment project. The project was initially planned as a conventional grey/hard engineering river improvement project, but with the active involvement of university staff with knowledge of NORM as coordinators, members of the local public were brought into the planning process, their awareness of the river environment changed, and the plan was transformed into the one based on NORM. A physical model of the river was created to share among stakeholders, and the riverside was designed with consideration for the restoration of desirable aesthetic features, such as the river's natural riffle-pool sequence. The revitalized river is now used for environmental education, and local citizens are engaged in maintenance and management, such as mowing the grass, toward which the maintenance budget is primarily allocated. Thanks to the participation of citizens from the beginning, a sustainable river beloved by local residents was created. Furthermore, the biodiversity of the river habitat has greatly increased since restoration, contributing to conservation goals (Figs. 11.2 and 11.3).

11.3.2 Disaster Recovery Projects and Nature-Oriented River Management

For small- and medium-sized rivers in Japan, major improvements are often only carried out after the occurrence of a disaster. Thus, post-disaster river improvement is

Fig. 11.3 The river has been widened after the restoration (July 25, 2015), and access to the waterfront has been improved, resulting in an enhanced wildlife habitat, and a significant increase in the number of people using the river recreationally (Kamisaigo River, photo by Dr. Hironori Hayashi)

an excellent opportunity for river restoration. However, owing to the need to implement post-disaster restoration works within a prompt timeframe, there have been cases where the environment has been neglected in favor of quick fixes. Therefore, two policies have been introduced to aid in the conservation of the river environment after disasters, when time is short and awareness of environmental conservation is low. The first policy is "Advisors for NORM" (Nakamura et al. 2013), wherein river restoration experts working in national research institutes (i.e., the Public Works Research Institute [PWRI] or National Institute for Land and Infrastructure Management [NILIM]) are dispatched at the request of the prefecture to the disaster-affected sites to provide advice on environmental considerations in river restoration. Experts may provide advice on more than one occasion, as needed. This scheme has produced the best restoration practices in many areas and has served as a reference for subsequent environmentally friendly flood control works. The second policy is to include the concept of NORM in the "Basic Policy for Disaster Recovery to Protect Beautiful Mountains and Rivers," a manual for disaster recovery projects for rivers (MLIT 2018). In more concrete terms, the contents and ideas of Point Book III for NORM, which is a compilation of knowledge gained from the experience of river restoration, were written into this basic manual to ensure the conservation of the river environment during disaster recovery projects. As a result, awareness of the need to conserve habitats and beautiful landscapes even in the short time available for river works following a disaster has increased significantly. However, many technical difficulties have yet to be addressed. For example, many of the small- and medium-sized rivers that suffer from disasters are mountain rivers, wherein the external force of water and sediment is great. In such cases, it can be a challenge to balance river restoration with flood prevention (Fig. 11.4).

Fig. 11.4 Example of a mountain river preserved by advisors for NORM (the Yamatsuki River, Miyazaki Prefecture)

11.3.3 Expansion into River Town Initiatives

With the development of NORM, there is growing interest in the diverse benefits of waterfront areas from the perspective of urban development. Especially in urban areas, there is an active movement to utilize the value of the waterfront in urban planning. The waterfront is a prized open space in the city, and its value as a place to relieve the stress of urban life, connect with nature, and enjoy recreation is increasing. These developments, which focus on the multifunctionality of the waterfront, can be called blue-green infrastructure (BGI). In Japan, MLIT is promoting BGI through a scheme known as River Town Initiatives (RTI). Once a project is designated an RTI, river administrators of the government or prefectural government provide support to the city government for planning and facility development.

The Doh-Tombori River, which runs through the central part of Osaka, was artificially created as a canal in the seventeenth century and its waterfront area has been bustling since the Edo period. However, in recent years, owing to the installation of high concrete blocks as a revetment and the deterioration of water quality, the river has become culturally devalued, with many riverside businesses choosing to set up their storefronts facing away from the river. At the beginning of the twenty-first century, the Osaka City Government set the goal of creating "Water Capital Osaka." In 2004, a promenade called "Tombori River Walk" was constructed, which was later extended with the addition of a boat landing. Consequently, many businesses that previously faced away from the river gradually began to set up their entrances on the riverside. In the future, it is expected that the town will be as lively as it was before (Nakai 2017).

The Tama River, which flows through the center of Tokyo and has a population of 4 million, was called a "dead river" in the 1970s because of its severe water pollution. However, its water quality has improved dramatically since then owing to the development of better sewage systems. The river naturally shows an abundance of ecological dynamics, such as the supply of sediment from upstream. It is now

Fig. 11.5 Futako-Tamagawa region of the Tama River (Photo by Dr. Takaaki Sakamoto)

recovering its typical river landscape and diverse ecosystem and is used as a place of relaxation and connection to nature by the citizens of metropolitan Tokyo. The area around the Futako-Tamagawa Station on the banks of the Tama River has become a popular area for urban development that takes advantage of the natural capital of the Tama River. It can therefore be referred to as a GI type of urban development that makes good use of the functions of the waterfront (Fig. 11.5).

11.3.4 Watershed-Based Flood Management and Green Infrastructure

Severe flooding, due to climate change coupled with urbanization, has become an issue worldwide, and Japan is no exception. In 2018, heavy rains in western Japan exacerbated by Typhoon No. 7 and a rainy season front caused severe damage and resulted in the loss of more than 200 lives. In the 2019 typhoon in eastern Japan, the damage spread to Tokyo and northern Japan, and had the typhoon's course been slightly different, Tokyo could have been severely flooded. In recent years, large-scale floods have occurred annually, threatening existing flood prevention measures that have been in place for many years.

As already mentioned, MLIT in charge of flood countermeasures has begun to promote flood control policies that integrate flood basins, in response to a 2020 report (Panel on Infrastructure Development 2020) that called for WFM. The WFM is a policy to further accelerate measures, such as the improvement of levees and the construction and reinforcement of dams, in light of the increasing severity and frequency of water-related disasters caused by climate change. It also aims to implement water-related disaster countermeasures through the cooperation of all parties involved in the basin, from the catchment area (the area where rainwater flows into the river) to the inundation area (the area liable to flooding). Although flood control measures have been implemented in the entire basin, so far they have mainly focused on urban areas. While the objective of flood control in the past was to safely discharge floodwaters to the sea through river channels, inundation is now assumed to be possible, and efforts to minimize damage in the event of floods will be implemented through both soft and hard measures. In other words, it is important to optimize flood prevention for the entire basin based on the understanding that flood disasters will occur.

To promote WFM, it is necessary to think in the light of GI. The National Diet of Japan has requested that "WFM efforts should contribute to the formation of an ecological network by promoting the concept of GI that makes use of the diverse functions of the natural environment, and by actively conserving or restoring the functions of ecosystems that contribute to disaster risk reduction." The idea of contributing to the formation of an ecological network through disaster countermeasures is truly a GI concept and shows a strong will to improve the watershed environment through WFM policies.

The concept of ecological networks was first introduced to Japan from Europe in 1995. In 2008, the NSS included ecological networks in its plan for land management, and efforts to construct these networks are ongoing in various regions. However, as mentioned in Sect. 11.2.3, it has not been easy for river administrators who do not have the budget or authority for watershed management. Nowadays, these watershed-scale restorations will make significant progress in line with the progress of WFM, instead of focusing only on river channels. By utilizing WFM and various financing schemes as described in Sect. 11.2.3, the restoration of the river basin environment can be realized.

11.4 Concluding Remarks

In Japan, NbS are being implemented in a variety of ways under the title of green infrastructure (GI). As discussed earlier in this chapter, the focus of GI differs from country to country. Japan's version of GI was triggered by the Great East Japan earthquake, and is characterized by the fact that even today, disaster prevention, such as flood prevention in rivers and their basins, is a major driver. River restoration associated with flood control projects has been undertaken since before the term GI was coined, and there are many successful examples. Since the introduction of the concept of GI, efforts have become even more diverse, growing from mere nature restoration projects to more complex ones that aim to revitalize towns. The methods of financing GI are evolving from regular public projects to a variety of projects utilizing private or personal funds. Japan is transitioning from a growing to a mature society; rather than focusing on economic development, it aims to enrich the lives of its people. Japan's experience, introduced in this chapter, offers one vision of how GI efforts can guide river restoration in a way that prevents disaster while preserving beauty and ecological integrity.

Acknowledgments This work was partially supported by the collaborative research project of the International Research Institute of Disaster Science (IRIDeS), Tohoku University.

References

CBD (2019) Global Biodiversity Outlook 5. https://www.cbd.int/gbo5. Accessed 28 Dec 2021

Conservation International (2021) Biodiversity hotspots. https://www.conservation.org/priorities/biodiversity-hotspots. Accessed 28 Dec 2021

EPA (2019) What is green infrastructure? https://www.epa.gov/green-infrastructure/what-green-infrastructure. Accessed 28 Dec 2021

EU (2013) The EU strategy on green infrastructure. https://ec.europa.eu/environment/nature/ecosystems/strategy/index_en.htm. Accessed 28 Dec 2021

Hayashi H, Shimatani Y, Hattori M (2016) A practice of stream restoration project in urban area and its evaluation. In: Proceedings of the 11th international symposium on ecohydraulics, Melbourne, Australia, 7–12. https://asnevents.s3.amazonaws.com/Abstrakt-FullPaper/26011/160115KamisaigoRiver_Final.pdf. Accessed 28 Dec 2021

IPBES (2019) Global assessment report on biodiversity and ecosystem services of the Intergovernmental Science-Policy Platform on Biodiversity and Ecosystem Services. Brondizio ES, Settele J, Díaz S, Ngo HT (eds). IPBES secretariat, Bonn, Germany. 1148 pages. https://doi.org/10.5281/zenodo.3831673

IUCN (2021a) The IUCN Red List of Threatened Species. [Version 2021a-3]. https://www.iucnredlist.org. Accessed 15 Jan 2021a

IUCN (2021b) Nature-based solutions. https://www.iucn.org/theme/nature-based-solutions. Accessed 28 Dec 2021b

Manabe M (2021) Juchu ni kakasenai shin-kyoyo green infra (Green infrastructure as new knowledge essential for winning new contracts). Nikkei Construction 9.27, p 33. https://xtech.nikkei.com/atcl/nxt/mag/ncr/18/00142/092100001/. Accessed 28 Dec 2021

MLIT (2006) Tashisen kawa zukuri kihon shishin (Basic principles for nature-oriented river management). https://www.mlit.go.jp/kisha/kisha06/05/051013/02.pdf. Accessed 28 Dec 2021

MLIT (2010) Chusyo kasen ni kansuru kado keikaku no gijutsu kijun (Technical standards for river restoration for small and medium-sized rivers). https://www.mlit.go.jp/river/shishin_guideline/kankyo/tashizen/pdf/gijyutsukijyun.pdf. Accessed 28 Dec 2021

MLIT (2015a) Kokudo keisei keikaku (National Spatial Strategy). https://www.mlit.go.jp/common/001100233.pdf. Accessed 28 Dec 2021

MLIT (2015b) Dai 4 ji syakai shihon seibi juten keikaku (Fourth priority plan for social infrastructure development). https://www.mlit.go.jp/sogoseisaku/point/content/001406606.pdf. Accessed 28 Dec 2021

MLIT (2018) Utsukushii sanga wo mamoru saigai hukyu kihon hoshin (Basic policy for disaster recovery to protect beautiful mountains and rivers). https://www.mlit.go.jp/river/shishin_guideline/bousai/saigai/measures-saigai/index.html. Accessed 28 Dec 2021

MLIT (2019) Green infra suishin senryaku (Green Infrastructure Promotion Strategy). https://www.mlit.go.jp/common/001297373.pdf. Accessed 28 Dec 2021

MLIT (2021a) Green infra kanmin renkei platform (Green Infrastructure Public-Private Partnership Platform). https://gi-platform.com/. Accessed 28 Dec 2021a

MLIT (2021b) Dai 5 ji syakai shihon seibi juten keikaku (Fifth priority plan for social infrastructure development). https://www.mlit.go.jp/sogoseisaku/point/content/001406599.pdf. Accessed 28 Dec 2021b

Nakai T (2017) Kanmin Renkei ni yoru Mizube Kukan no Nigiwai Soshutsu - Osaka Dotombori-Gawa no Jirei yori (Creating a lively waterfront space through public-private partnerships: A case study of the Dotonbori River in Osaka). 21st Century Wakayama 87:32–36. http://www.wsk.or.jp/book/87/08.pdf. Accessed 16 Jan 2022

Nakamura K, Tockner K, Amano K (2006) River and wetland restoration: lessons from japan. Bioscience 56(5):419–429. https://doi.org/10.1641/0006-3568(2006)056[0419:RAWRLF]2.0.CO;2

Nakamura K, Haga M, Iwami Y (2013) River restoration advisor for post-disaster river works in Japan. In: The 5th European River Restoration Center, ERRC2013. https://www.researchgate.net/publication/344711277_River_restoration_advisor_for_post-disaster_river_works_in_Japan. Accessed 28 Dec 2021

Panel on Infrastructure Development (2020) Kiko Hendo wo humaeta Mizu Saigai Taisaku no Arikata (Water-related disaster countermeasures in light of climate change). https://www.mlit.go.jp/river/shinngikai_blog/shaseishin/kasenbunkakai/shouiinkai/kikouhendou_suigai/pdf/03_honbun.pdf. Accessed 16 Jan 2021

Research Group on Nature-oriented River Management (2011) Tashizen kawa zukuri point book (Point book III for nature-oriented river management)

Science Council of Japan (2014) Tegen fukko/kokudo-kyojinka ni okeru seitaikei Infrastructure katsuyo no susume (Recommendations for the use of ecosystem infrastructures in reconstruction and national land resilience). https://www.scj.go.jp/ja/info/kohyo/pdf/kohyo-22-t199-2.pdf. Accessed 28 Dec 2021

The World Bank (2020) Forest area (% of land area). https://data.worldbank.org/indicator/AG.LND.FRST.ZS. Accessed 28 Dec 2021

Waley PT, Aberg EU (2011) Finding space for flowing water in Japan's densely populated landscapes. Environ Plan a: Int J Urban Reg Res 43(10):2321–2336. https://doi.org/10.1068/a4489

Chapter 12
Future Perspectives of Financing Investment in Disaster Risk Reduction and Climate Change Adaptation

Mikio Ishiwatari and Daisuke Sasaki

Abstract It is widely recognized that investment is crucial to mitigating disaster damage and adapting to the effects of climate change. This point was emphasized in the Sendai Framework for Disaster Risk Reduction and successive agreements on climate change. The preceding chapters of this book identified various issues that need to be resolved and proposed policies and approaches accordingly. This chapter synthesizes the major findings and proposes innovative approaches to attain sustainable progress toward a resilient future in the Asian region. There are research gaps in many areas of investment in disaster risk reduction. There is a need to develop the estimating methods and the database of investments. The activities that constitute disaster management phases need to be well-coordinated and further developed. The trend in the increasingly adverse allocation of investment between ex-ante before disaster occurrence and response and recovery following disasters needs to be reversed. Since policy- and decision-makers must justify the allocation of budget among various priority activities, the methods to analyze the benefits of disaster risk reduction and climate change adaptation need to be improved. To overcome the paucity of finance, further research should be aimed at developing innovative approaches, such as green infrastructure, partnerships with the private sector, and burden-sharing mechanisms.

Keywords Continuity · Disaster recovery · Financial gap · Resilient infrastructure · Sendai Framework for Disaster Risk Reduction

---

M. Ishiwatari (✉)
Graduate School of Frontier Sciences, The University of Tokyo, Chiba, Japan
e-mail: ishiwatari.mikio@jica.go.jp

D. Sasaki
International Research Institute of Disaster Science (IRIDeS), Tohoku University, Sendai, Miyagi, Japan
e-mail: dsasaki@irides.tohoku.ac.jp

M. Ishiwatari and D. Sasaki (eds.), *Financing Investment in Disaster Risk Reduction and Climate Change Adaptation*, Disaster Risk Reduction,
https://doi.org/10.1007/978-981-19-2924-3_12

12.1 Introduction

The growing recognition of the importance of investing in disaster risk reduction (DRR) and climate change adaptation (CCA) is reflected in the emphasis placed on this point in the Sendai Framework for DRR and agreements on climate change (Ishiwatari 2021a; United Nations 2019; UNISDR 2015). The preceding chapters presented an overview of the investments in these areas, raised various issues that need to be resolved, and identified the opportunities to improve the systems for reducing disaster risks. This chapter derives and synthesizes the major findings of the previous chapters. Further, it proposes investment policies and innovative approaches to achieve sustainable progress toward a resilient future in the Asian region. The chapter concludes by indicating the directions that future research may follow to fill the observed research gaps.

12.2 Amount of Investment in DRR and CCA

This section summates the levels of current and future investments in CCA and DRR observed in the previous chapters of this book. Figure 12.1 shows the current level of the estimated annual expenditure on protection against floods in Asia, CCA in the world, and the future needs for these purposes. The need for investment in flood protection in Asia by 2030 is projected to be approximately 100 billion USD, that is, nearly thrice the investment observed in 2015. CCA will require an investment of at least 100 million USD in 2030 (Fujikura 2022). Some of the amounts to be invested in flood protection may have been included in the projected investment in CCA. Further, there is no unanimity or exactitude in the estimated amounts because the authors have employed various methods of estimation.

Figure 12.2 shows the amount of finance flowing from industrialized countries to developing countries for DRR, humanitarian assistance, and CCA. There are bound to be overlaps in these amounts, but these cannot be identified because the data have been drawn from various sources. The amount of humanitarian assistance, which includes the response to man-made and natural disasters, is similar to CCA and larger than DRR. Investment in the ex-ante measures of DRR, that is, preparing for and mitigating disaster risks, accounts for less than 10% of DRR financing, humanitarian assistance, and CCA financing. Financing CCA aimed at strengthening public resilience in coastal regions accounts for 5–10% of total CCA financing.

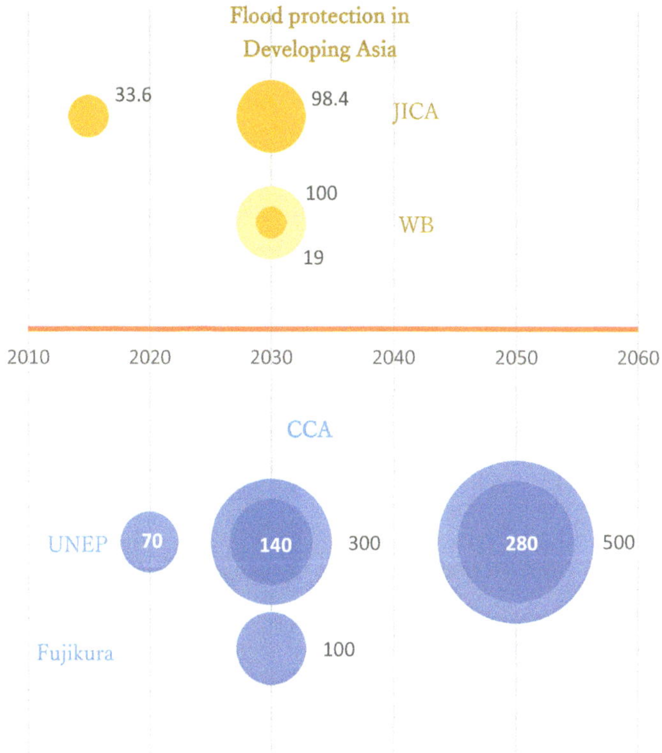

Fig. 12.1 Estimated annual needs for flood protection in Asia and CCA in the world. Dark color circles indicate lower estimates and light circles higher estimates. Numbers are billions of USD. *Source* Authors' elaboration after Ishiwatari and Sasaki (2020), Rozenberg and Fay (2019), UNEP (2021), and Fujikura (2022)

12.3 Issues to Be Resolved, Required Approaches, and Research Gaps

The issues, approaches, and research gaps identified in the previous chapters are discussed in this section. These include the use of unaccredited databases, disconnect among phases of disaster management, difficulties in justifying investment, and developing innovative measures (Fig. 12.3).

12.3.1 Data Definition and Accurate Estimation of Financing DRR and CCA

Building a comprehensive view of CCA financing is challenging because the estimating methods and database of CCA financing have not been sufficiently developed.

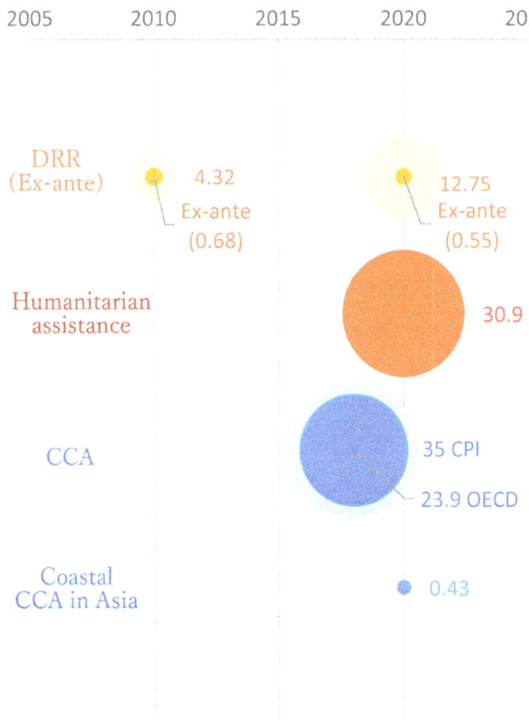

Fig. 12.2 Finance to DRR, Humanitarian Assistance, and CCA. Numbers are billions of USD. Dark color circles in DRR indicate ex-ante amounts in DRR; the dark circle in CCA indicates Organisation for Economic Co-operation and Development (OECD) estimate, and the light color CPI estimate. *Source* Authors' elaboration after Kellett and Caravani (2013), UNDRR (2021), Development Initiatives (2021), Buchner et al. (2019), CPI (2019), OECD (2020), and Shiiba (2022)

In Chap. 2, Fujikura (2022) argues that the reported amount of CCA financing could be overestimated because of the lack of clarity in estimation methods. This argument is substantiated by Shinohara's report in Chap. 5 that, in the Philippines, the proportion of CCA financing by local government units dominated the infrastructure projects, such as highways. The reason is that the organizations in the Philippines categorize the entire estimated project costs as CCA costs if the project includes CCA activities, even as a part of the total project. For example, if a highway project includes the works of disaster prevention, the cost of the entire project could be categorized as an investment in CCA (Shiohara 2022).

Shiiba (2022), in Chap. 7, examines the international financing of CCA for coastal resilience projects. Since the database of coastal resilience projects is not available, Shiiba collected the data on projects from multilateral development banks (MDB).

The Intergovernmental Panel on Climate Change (IPCC 2014) defined adaptation as "the process of adjustment to actual or expected climate and its effects."

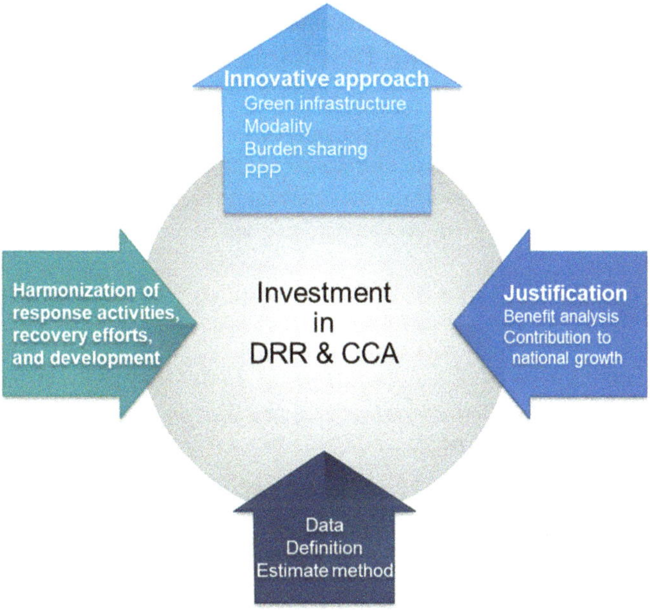

Fig. 12.3 Required approaches identified in the book. *Source* Authors' elaboration

Theoretically, incremental costs that will be incurred in adapting to the effects of climate change should be counted as CCA costs. Furthermore, MDB agrees with the methodology of estimating CCA financing (MDB 2019). CCA projects must be based on clear identification and evidence-backed assessment of the vulnerability of the activities to be undertaken in the project areas (Barbara 2019). The CCA activities should be distinguished from the standard development components of a project. CCA activities must be aimed clearly at improving social resilience to climate change. While the methodology is clear, without much distinction, the banks are unable to accurately count the incremental cost of adapting to climate change.

Distinguishing the elements of adaptation finance from development operations is another challenge. The two are interconnected, and incremental costs of CCA activities are usually not shown separately in the total financial requirements of projects. For example, it is difficult to segregate the incremental costs of CCA from drainage projects. It is necessary to project the increase in rainfall and the resultant increase in the volume of flood waters to facilitate the projection of the increase in the cost of expanding the drainage facility to avoid the adverse effects of climate change. Further, this cost should be separated from the cost of managing floods. It is necessary to develop detailed methodologies for this purpose. Moreover, since CCA relates to several areas, such as DRR, water, and agriculture, collecting and financing data from scattered data sources require huge efforts (Timilsina 2021).

12.4 Harmonization of Pre-disaster, Response Activities, Recovery Efforts, and Development

Various international and national organizations are involved in financing and coordination of the successive phases of disaster management, which are pre-disaster, response activities, recovery effort, and development. The various organizations involved lead to poor coordination among these activities and an imbalance in their funding.

12.4.1 Imbalance in the Allocation of Assistance Funding Among the Pre-disaster and the Response and Recovery Phases

Countries and assistance agencies should emphasize taking proactive measures before a disaster strikes to prevent or minimize the damage caused by the disaster. As UNISDR (2015) stressed, investment in structural and non-structural measures to prevent or reduce the risk of disaster is essential and cost-effective in saving lives and preventing and reducing losses.

However, as Ishiwatari and Sasaki (2022) stated in Chap. 1, only limited amounts of international assistance for natural disasters are allocated to the ex ante measures, and this situation is worsening. Financing for ex ante measures declined from $680 million/year in 1991–2010 to $550 million/year in 2011–2020, and the share of DDR assistance in this finance declined from 12.7% to 4.1% (Kellett and Caravani 2013; UNDRR 2021).

The declining trend in ex ante measures indicates the difficulties in securing funds for such measures before a disaster occurs. DRR facilities, such as dykes and early warning systems, can mitigate human and economic losses, but cannot provide tangible benefits unlike other infrastructures, such as roads and water supply. Most DRR facilities are useful only when disasters strike, but they do not provide any services to the general public in normal times (Ishiwatari and Sasaki 2021a). Large structures, such as dams and diversion channels may take years to construct and the benefits from such projects are not available to the public until they are completed. The execution of such projects requires large financial resources and technical expertise, but both are limited in developing countries.

In contrast to building DRR facilities, when a disaster strikes, governments and assistance agencies naturally respond to the needs of the affected people that are evident. The victims need food, water, other relief supplies, shelter, and sustained recovery efforts to restore livelihoods, houses, and communities. The damaged infrastructure, such as communication, electricity, and other public utilities, needs to be repaired without delay to be useful in response operations and rehabilitation programs.

The High-level Experts and Leaders Panel on Water and Disasters (HELP) was established in 2007 to assist the international community, governments, and stakeholders in mobilizing political will and resources to deal with disasters. In its global report of 2019, HELP stressed striking a rational balance between the proportions of finance allocated to preparedness and mitigation on the one hand, and emergency response and rehabilitation on the other.

12.4.2 Barriers to Accomplishing Continuity from Response to Recovery and Development

Besides the imbalanced allocation of funds among ex ante activities and response and recovery activity phases, there is a lack of coordination in the implementation of these phases and with the development that follows recovery. There are barriers to accomplishing continuity in the range of activities from disaster response to recovery and development.

In Chap. 8, Katsube (2022) considered four recent major disasters in Asia Pacific Region and puts the estimated need for funds for early recovery at 0.04–0.06% of the total funds needed for humanitarian assistance. The insignificance of this amount indicates that early recovery is not recognized as a life-saving activity, that is, the finance earmarked for humanitarian assistance is not readily available for recovery efforts.

In Chap. 9, Jibiki et al. (2022) argued that there was a disconnect between recovery efforts and development activities. The reason for this disconnect in the cases of the Philippines and Indonesia is the limited coordination between the recovery efforts made by the offices of disaster management and climate change. Jibiki et al. pointed out that investment in DRR and CCA is not adequately covered by the laws, ordinances, and guidelines regarding post-disaster needs assessment (PDNA). While PDNA mainly estimates the need for rehabilitating facilities and economic activities in a wide range of sectors, it also unnecessarily covers the need for making society more resilient to climate change and disasters. Policy research to develop approaches to connect recovery efforts with development is necessary.

12.5 Justification of Investment in DRR and CCA

It is necessary to justify the demanded quantum of investment in DRR and CCA at the national level and project scale (Ishiwatari 2019, 2021b; Kawasaki and Rhyner 2018). Moreover, the factors that affect the quantum of finance should be analyzed to facilitate the strengthening of investment policies.

12.5.1 National Level

To gain the support of policymakers for investment in DRR and CCA, the contribution of these activities to national growth must be explained to them. In Chap. 3, Sasaki and Ishiwatari (2022) examined new beneficial concepts of DRR and CCA proposed by The World Bank and Asian Development Bank (ADB). The new concept of the triple dividend of resilience consists of (i) avoiding disaster losses, (ii) stimulating economic activities, and (iii) providing co-benefits in socio-economy and environment. ADB also emphasized the importance of women-focused investment. Some academic papers have supported these concepts. However, further studies are needed to strengthen these concepts theoretically.

Few studies have empirically analyzed the relationship between investment and damage at the national level (Ishiwatari and Sasaki 2021a). Tsukahara and Kachi (2016) estimated that Japan benefits by several trillion JPY (tens of billions of USD) annually from investment in flood protection. Ishiwatari and Sasaki (2021b) adopted this method and found that investments in flood protection in the People's Republic of China (PRC) and the Philippines are cost-effective.

12.5.2 Project Scale

It is necessary for organizations that implement the DRR and CCA projects to show their cost-effectiveness on a project scale. Various studies have found that DRR measures are cost-effective. For example, the cost–benefit ratios of mitigation programs in the United States vary from 4:1 to 11:1. The mitigation programs cover the countermeasures, such as adopting building codes, retrofitting infrastructure, and projects for preventing riverine and coastal flooding, damage by hurricanes, earthquakes, and fires implemented by spending the grants from the federal agencies' grant programs (Multi-Hazard Mitigation Council 2019). Davlasheridze et al. (2017) examined measures against hurricanes in the United States and found that a marginal return from long-term mitigation projects for pre-empting hurricane damage is almost twice that of the marginal return from post-hurricane recovery programs.

As Sakamoto et al. (2022) pointed out in Chap. 11, while there are studies that examined the cost-effectiveness of the DRR projects, there is a need to further refine the methods of estimating benefits arising from DRR. Since 1970, the Japanese government has developed methods to estimate the benefits of flood protection. However, the methods do not estimate the benefits of saving human lives and minimizing long-term economic losses because of the complexity of these beneficial effects.

12.5.3 Resilient Infrastructure

In Chap. 6, Mizutani (2022) stressed the importance of maintenance to keep infrastructure resilient against disasters. Without proper maintenance, structures can deteriorate over time and become vulnerable to disasters. Further studies need to develop methods to optimize the lifecycle costs of disaster-management infrastructure.

12.5.4 Factors Affecting Investment Needs

Understanding factors affecting the need of investment in DRR is crucial for strengthening investment policies. Pelupessy et al. (2022) studied people's perception of investing in flood protection in Jakarta and found that community members would have a heightened sense of resilience as developing a sense of place attachment and collective efficacy. This points toward the need for research to assess people's perceptions of investment in DDR in other countries at the community and national levels.

12.6 Innovative Measures of Investment in DRR and CCA

There is a need to develop innovative means of financing to increase the investment in DRR and CCA projects and the efficiency of their implementation. Stakeholders should share roles and responsibilities in financing projects in various countries based on the characteristics and history of each country. While the public sector meets the costs of DRR and CCA, the private sector and local communities could share some responsibility and costs.

12.6.1 Issues Found

The world could not achieve the target of investing 100 billion USD annually in climate financing in 2020 and 2021 to which it had agreed at the 2009 United Nations Climate Change Conference in Copenhagen (Timperley 2021). However, industrialized countries are expanding finance to CCA in developing countries to reach the agreed target.

In Chap. 2, Fujikura (2022) pointed out the importance of designing new modalities to finance CCA. Furthermore, new modalities should be developed to tackle the challenge of involving the private sector in financing CCA. The topics discussed in the other chapters were some of the new financing modalities for CCA, green infrastructure, local governments' contribution, and the partnership between public and private sectors.

In Chap. 10, Nakamura (2022) examined the Japanese approaches to green infrastructure and stressed that green infrastructure could produce multiple benefits. These include, besides DRR or CCA, benefits in a wide range of areas, such as climate change mitigation, environmental conservation, and stimulating the local economies.

National and local governments share the responsibilities in creating DRR and CCA facilities and their respective financial burden of investments. Local governments in the Philippines play crucial roles in promoting CCA. In Chap. 5, Shiohara (2022) analyzed their budgets for CCA. While local governments have the primary responsibility for DRR, national governments provide the local governments with financial support for developing DRR facilities as the basic infrastructure.

12.6.2 Research Gaps

It was found that there are research gaps in developing the approaches to mobilizing finance from various stakeholders. There is no common agreement across societies over the sharing of burden among national governments, local governments, local communities, and the private sector. Further, for proposing new policies of financing, empirical research into the following issues is necessary.

In Japan, within the context of decentralization, cost-sharing between national and local governments for flood protection projects became a controversial issue during the late 2000s. Local governments cover one-third of the costs for the national projects of flood protection from the viewpoint of sharing expenses by beneficiaries. The national governments insisted that the national government should promote works for river basins, covering multiple prefectures through coordinating local governments and envisage projects to secure balanced safety throughout the country. Local governments required more initiatives in deciding on flood protection projects rather than just following the national government's decisions. It is because the national government did not disclose the detailed information on projects, such as the breakdowns and backgrounds of costs, nor did it properly consider the financial and societal situations of local governments for project sites and schedules. Furthermore, local governments require devolving financial resources and responsibilities of flood protection from the national governments to local governments.

Local communities have certain resources which they contribute as their share of investment in DRR and CCA. For example, community-based organizations in the Netherlands and Japan have historically been actively involved in disaster management by taking flood management measures to protect themselves (Ishiwatari and Surjan 2019). Water boards are the oldest democratic organizations in the Netherlands. They began water management including flood protection in their communities during the thirteenth century. At their peak, several thousand boards were active and currently, 21 boards are functioning (Van Stokkom et al. 2005). They can collect revenues in the form of the water board charges and the pollution levy. The water boards are expected to play an important role in adapting local water systems to meet the impacts caused by climate change (Kamperman and Biesbroek 2017).

In Japan, the local communities of farmers became autonomous in the middle or early modern period and began developing facilities to protect their paddy fields and houses against flooding (Ishiwatari 2012). In the early seventeenth century, the local communities of the Nobi plain constructed dykes around the communities to keep flood waters out of their communities (Baba and Ito 2018).

While investment in DRR and CCA is regarded as unprofitable and may not attract investment from the private sector (Timilsina 2021), the private sector can play an important role in these activities (Mavrodieva et al. 2019). The government can encourage the private sector to invest in DRR and CCA by granting tax rebates, relaxing certain regulations, and providing guidance in constructing the facilities for mitigating disaster damage (Ishiwatari 2016). Khattri et al. (2010) identified three reasons why the private sector might invest in CCA and DRR: (i) to maintain business continuity by protecting their operations from disasters and the effects of climate change, (ii) if they are stakeholders in developing the infrastructure and providing technical solutions, (iii) if CCA and DRR projects provide a market for their products.

12.7 Conclusion

Though the Sendai Framework for DRR and international agreements on climate change urge governments and other entities to invest in DRR and CCA, several issues that hamper such investment have been identified in this book. For example, there are research gaps in fundamental areas. Even estimating the current scale of investments is a challenge. Further, estimating the projected need for investments is a complex task. Therefore, there is a need to develop estimating methods and databases.

The crucial issues of connecting the activities within disaster management phases and with the overall development have been discussed (Gómez and Kawaguchi 2018). It is widely recognized that investment in disaster mitigation and preparedness is cost-effective (UNISDR 2015). However, the complexity of balancing investment between ex ante and response and recovery is increasing. Policy- and decision-makers need justification for allocating budgets for DRR and CCA. It is necessary to develop the methods to analyze the benefits that are to be derived from individual DRR and CCA projects and sum-total of all project-level benefits at the national level. To bridge financial gaps, further research is needed to develop innovative approaches to green infrastructure, partnership between public and private sectors, and burden-sharing mechanisms.

As mentioned above, to attract increased investment in DRR and CCA, several major issues need to be resolved. The issues are (i) the definition of data and accurate estimation of the necessary quantum of finance for executing DRR projects and CCA activities; (ii) ensuring harmony among response activities, recovery efforts, and development; and (iii) devising methods to weigh the costs and benefits of DRR and CCA to justify the investment. Academic research can, doubtlessly, play a key role in finding clues that would lead to the resolution of these issues. The previous chapters of this book have shed light on some innovative approaches and measures that could, possibly, be the solutions to the above issues. The authors note that exhaustive research on innovative approaches is needed. In the end, we hope that this book can contribute to bringing DRR and CCA recognition as profitable avenues of investment.

Acknowledgments This study was supported by the collaborative research project of the International Research Institute of Disaster Science, Tohoku University. This study was also partially supported by the Japan Society for the Promotion of Science KAKENHI, Grant Number 21H03680.

References

Baba S, Ito M (2018) Historical background of the countermeasures against floods along the upstream area of the Kiso River: focusing on the embankment construction for the Spur Dike (in Japanese). J Environ Inform 32:239–244

Barbara B et al (2019) Global Landscape of Climate Finance. Climate Policy Initiative https://www.climatepolicyinitiative.org/publication/global-landscape-of-climate-finance-2019/

Buchner B et al (2019) Global landscape of climate finance 2019. Climate Policy Initiative, London. https://climatepolicyinitiative.org/publication/global-climate-finance-2019/

CPI (2019) Global landscape of climate finance 2019. Climate Policy Initiative, London

Davlasheridze M et al (2017) The effects of adaptation measures on hurricane induced property losses: which FEMA investments have the highest returns? J Environ Econ Manage 81:93–114

Development Initiatives (2021) Global Humanitarian Assistance Report 2021. https://devinit.org/resources/global-humanitarian-assistance-report-2021/

Fujikura R (2022) Financing in climate change adaptation. In: Ishiwatari M, Sasaki D (eds) Financing investment in disaster risk reduction and climate change adaptation: Asian perspectives. Springer, Tokyo

Gómez OA, Kawaguchi C (2018) A theory for the continuum: multiple approaches to humanitarian crises management. In: Hanatani A, Gómez OA, Kawaguchi C (eds) Crisis management beyond the humanitarian-development nexus. Routledge, pp 15–25

High-level Experts and Leaders Panel on Water and Disasters (HELP) (2019) Principles on investment and financing for water-related disaster risk reduction. https://www.wateranddisaster.org/cms310261/wp-content/uploads/2019/07/HELP-Principles-Full-Final-Printing.pdf

IPCC (2014) Annex II: glossary. In: Climate change 2014: synthesis report. contribution of working groups I, II and III to the Fifth assessment report of the Intergovernmental Panel on Climate Change. IPCC, Geneva, Switzerland

Ishiwatari M (2016) What are crucial issues in promoting an integrated approach for flood risk management in urban areas? JSIJ 6(1):15–26

Ishiwatari M (2019) Investing in disaster risk reduction: scale and effect of investment in flood protection in Asia. UNDRR, Geneva

Ishiwatari M (2021b) Effectiveness of investing in flood protection in metropolitan areas: lessons from 2019 Typhoon Hagibis in Japan. Int J Disaster Resil Built Environ 13(1):89–98

Ishiwatari M, Sasaki D (2021a) Investing in flood protection in Asia: an empirical study focusing on the relationship between investment and damage. PDISAS 100197. https://doi.org/10.1016/j.pdisas.2021a.100197

Ishiwatari M, Surjan A (2019) Good enough today is not enough tomorrow: challenges of increasing investments in disaster risk reduction and climate change adaptation. Prog Disaster Sci 1:100007
Ishiwatari M, Sasaki D (2020) Bridging the gaps in infrastructure investment for flood protection in Asia. JICA Research Institute, Tokyo
Ishiwatari M, Sasaki D (2021b) Investments in flood protection: trends in flood damage and protection in growing Asian economies. JICA Research Institute, Tokyo
Ishiwatari M, Sasaki D (2022) Financing investment in disaster risk reduction and climate change adaptation: opportunities and challenges. In: Ishiwatari M, Sasaki D (eds) Financing investment in disaster risk reduction and climate change adaptation: Asian perspectives. Springer, Tokyo
Ishiwatari M (2012) Government roles in community-based disaster risk reduction. In: Shaw R (ed) Community-based disaster risk reduction (community, environment and disaster risk management, vol 10). Emerald Group Publishing Limited, Bingley
Ishiwatari M (2021a) Disaster risk reduction. In: Lackner M, Sajjadi B, Chen WY (eds) Handbook of climate change mitigation and adaptation. Springer, New York. https://doi.org/10.1007/978-1-4614-6431-0_147-1
Jibiki Y et al (2022) Is the post disaster needs assessment important to investment in disaster risk reduction and climate change adaptation? Lessons from Indonesia and the Philippines based on literature review. In: Ishiwatari M, Sasaki D (eds) Financing investment in disaster risk reduction and climate change adaptation: Asian perspectives. Springer, Tokyo
Kamperman H, Biesbroek R (2017) Measuring progress on climate change adaptation policy by Dutch Water Boards. Water Resour Manag 31:4557–4570. https://doi.org/10.1007/s11269-017-1765-8
Katsube T (2022) Early recovery in humanitarian response-in cases of four disasters in Asia Pacific Region. In: Ishiwatari M, Sasaki D (eds) Financing investment in disaster risk reduction and climate change adaptation: Asian Perspectives. Springer, Tokyo
Kawasaki A, Rhyner J (2018) Investing in disaster risk reduction for resilience: roles of science, technology, and education. J Disaster Res 13(7):1181–1186
Kellett J, Caravani A (2013) Financing disaster risk reduction: a 20 year story of international aid. GFDRR and ODI
Khattri A et al (2010) Opportunities for private sector engagement in urban climate change resilience building. Report by Intellectual Capital Advisory Services Pvt. Ltd. (Intellecap) with the support of the Asian Cities Climate Change Resilience Network (ACCCRN) and the Rockefeller Foundation. Intellecap, Hyderabad, India and the Rockefeller Foundation, Bangkok
Mavrodieva AV et al (2019) Governmental incentivization for SMEs' engagement in disaster resilience in Southeast Asia. IJSRM 1(1):32–50
Mizutani D (2022) Financing Infrastructure of DRR: Possibility of integrating investments in maintenance and DRR. In: Ishiwatari M, Sasaki D (eds) Financing investment in disaster risk reduction and climate change adaptation: Asian Perspectives. Springer, Tokyo
Multi-Hazard Mitigation Council (2019) Natural hazard mitigation saves: 2019 report. National Institute of Building Sciences, Washington, DC
Multilateral Development Banks (MDB). (2019) Joint report on multilateral development banks' climate finance. 2018. http://www.ebrd.com/2018-joint-report-on-mdbs-climate-finance
Nakamura K (2022) Nature-based solution in Japan: practice of river restoration. In: Ishiwatari M, Sasaki D (eds) Financing investment in disaster risk reduction and climate change adaptation: Asian perspectives. Springer, Tokyo
Organisation for Economic Co-operation and Development (OECD) (2020), Climate finance provided and mobilised by developed countries in 2013–2018. OECD Publishing, Paris. https://doi.org/10.1787/f0773d55-en
Pelupessy D et al (2022) Exploring people's perception of disaster risk reduction investment for flood management: a case from Jakarta floods in Indonesia. In: Ishiwatari M, Sasaki D (eds) Financing investment in disaster risk reduction and climate change adaptation: Asian Perspectives. Springer, Tokyo

Rozenberg J, Fay M (eds) (2019) Beyond the gap: how countries can afford the infrastructure they need while protecting the planet. Sustainable Infrastructure Series. World Bank, Washington, DC
Sakamoto M et al (2022) Estimating the benefits of flood protection: Reviewing the transition of the cost-benefit analysis manual in Japan and the implementation status of the international flood protection project. In: Ishiwatari M, Sasaki D (eds) Financing investment in disaster risk reduction and climate change adaptation: Asian Perspectives. Springer, Tokyo
Sasaki D, Ishiwatari M (2022) Recent trends in disaster risk reduction investments: a literature review. In: Ishiwatari M, Sasaki D (eds) Financing investment in disaster risk reduction and climate change adaptation: Asian Perspectives. Springer, Tokyo
Shiiba N (2022) Financing climate-resilient coasts: tracking multilateral aid for ocean and coastal adaptation to climate change in Asia–Pacific. In: Ishiwatari M, Sasaki D (eds) Financing investment in disaster risk reduction and climate change adaptation: Asian Perspectives. Springer, Tokyo
Shiohara K (2022) Factors influencing climate change adaptation investment by local government units in the Philippines. In: Ishiwatari M, Sasaki D (eds) Financing investment in disaster risk reduction and climate change adaptation: Asian Perspectives. Springer, Tokyo
Timilsina GR (2021) Financing climate change adaptation: International initiatives. Sustainability 13:6515. https://doi.org/10.3390/su13126515
Timperley J (2021) How to fix the broken promises of climate change. Nature 598:400–402
Tsukahara K, Kachi N (2016) Using data and statistics to explain investment effectiveness on flood protection. J Disaster Res 11(6):1238–1243. https://doi.org/10.20965/jdr.2016.p1238
UNDRR (2021) International cooperation in disaster risk reduction. United Nations Office for Disaster Risk Reduction, Geneva
UNISDR (2015) Sendai Framework for disaster risk reduction 2015–2030. UNISDR, Geneva
United Nations Environment Programme (UNEP) (2021) Adaptation gap report 2020. UNEP, Nairobi
United Nations (2019) Report of the secretary-general on the 2019 Climate Action Summit and the way forward in 2020. https://www.un.org/en/climatechange/assets/pdf/cas_report_11_dec.pdf
Van Stokkom HT et al (2005) Flood defense in the Netherlands: A new era, a new approach. Water Int 30(1):76–87

www.ingramcontent.com/pod-product-compliance
Ingram Content Group UK Ltd.
Pitfield, Milton Keynes, MK11 3LW, UK
UKHW022226051026
15624UKWH00002B/381
* 9 7 8 9 8 1 1 9 2 9 2 3 6 *